Hydraulic gates and valves

in free surface flow
and submerged outlets

Hydraulic gates and valves

in free surface flow
and submerged outlets

JACK LEWIN (Hon) DEng, CEng, FICE, FIMechE, FCIWEM

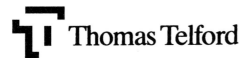

Published by Thomas Telford Publications, Thomas Telford Services Ltd, 1 Heron Quay, London E14 4JD

First published 1995

Distributors for Thomas Telford books are
USA: American Society of Civil Engineers, Publications Sales Department,
345 East 47th Street, New York, NY 10017-2398
Japan: Maruzen Co. Ltd, Book Department, 3–10 Nihonbashi 2-chome, Chuo-ku,
Tokyo 103
Australia: DA Books and Journals, 648 Whitehorse Road, Mitcham 3132, Victoria

A catalogue record for this book is available from the British Library

Classification
Availability: Unrestricted
Content: Guidance based on best current practice
Status: Author's analysis
Users: Hydraulic, Water and Civil Engineers, Designers

ISBN: 0 7277 2020 1

Typeset by Gray Publishing, Tunbridge Wells, Kent
Printed in Great Britain by Redwood Books, Trowbridge, Wiltshire

Contents

Foreword

Engineering is the marrying of academic knowledge to practical experience in order to carry a project to a successful conclusion. Hydraulic gates and valves form an important — often vital — part of reservoirs, barrages and river control structures. From extensive papers on hydraulic model studies and my own investigations it is apparent that there have in the past been defects in the design of gates or the water conveying system. In some cases, available knowledge has not been applied in practice because it was not easily accessible or widely disseminated.

This book was written to bridge the gap between theory and practice. I have attempted to identify poor design features in gates and point out some hydraulic misconceptions. I have provided an overview of the subject, recorded some experiences of designing gates and drawn attention to details which are important in achieving satisfactory operation.

Gates and valves control the flow of water. An understanding of how they interact with hydrodynamics is essential. The book therefore devotes several chapters to hydraulic considerations which affect gate design and operation. It does not include a detailed design guide or worked examples. One reason for this is that many aspects of gate design are now carried out using computers, making it difficult to adopt a specific design approach. (It also creates problems when checking gate contractors' calculations, and in many cases I have preferred to carry out independent computations.) There is also a limit on the time a practising engineer can spend writing a book.

The book is offered as a guide to engineers who work on projects which incorporate gates and valves and to engineers who design them. I hope they will find the information it contains useful. It may also benefit those responsible for the operation of river and reservoir control structures by identifying operational problems.

I am indebted to the engineers who read the manuscript and made valuable suggestions based on their own experience, particularly Mr Derek Wilden of Sir Alexander Gibb and Partners.

Jack Lewin, 1995

Acknowledgements

My thanks are due to my wife and daughter who struggled to translate the manuscript into a readable form.

Acknowledgement and thanks for permission to reproduce material from the author's papers, are made to IWEM for Figures 2.8, 2.16, 2.40, 2.41, 2.42, 2.44, 2.45, 2.48, 3.16, 6.2, 7.12, 10.10, 10.13, 10.14. Similarly to:

Bridon Ropes Ltd for Table 7.1.

British Hydromechanics Research Group, Cranfield for Figures 3.4, 3.5, 3.6, 3.7.

Computational Mechanics Centre, Southampton for Figures 2.11, 2.12, 2.13, 2.14, 2.15.

The Consorzio Venezia Nuova for Figures 2.33, 2.34.

Delft Hydraulics for Figure 17.2.

The Electricity Corporation of New Zealand, Waikato Hydro Group and Water Power and Dam Construction for Figures 11.8, 11.9, 11.10, 11.11, 11.12

Hydraulic Research, Wallingford for Figure 9.18.

The Institution of Civil Engineers for Figure 2.36.

The Institution of Mechanical Engineers for Figure 2.35.

Ishikawajima Harima Heavy Industries Co. Ltd for Figures 2.38, 2.43, 2.48 and the photograph in Figure 2.48.

Hans Künz gmbh for Figure 4.3.

Mannesmann Rexroth for Figure 6.5.

National Rivers Authority, Northumbria and Yorkshire Region for Figure 2.37.

Renold plc for Figure 7.17(c).

SKF (U.K.) Ltd for Figures 7.13, 7.15.

Voest Alpine gmbh for the photograph in Figure 3.18.

J.M. Voith gmbh for Figures 3.1, 3.2, 3.3, 3.8, 3.16, 3.17.

I am indebted to the engineers who scrutinised the manuscript and made useful suggestions, also to Sue Lamb for tracing the diagrams.

I owe much to Dr Paul Kolkman, formerly of Delft Hydraulics, for his generous permission to use material from his extensive papers, fundamental to an understanding of gate vibration.

1
Introduction

The control of rivers, canals and reservoirs requires weirs or appurtenances. From consideration of reliability and maintenance, the fixed weir is the preferred control structure. Similarly, the fixed-crest, free-overflow spillway is the most advantageous arrangement for reservoirs. Wherever weirs or fixed-crest overflow spillways cannot be accommodated, or where the backwater stages of a flood or variable river levels are unacceptable, a device which provides a moveable crest or a submerged variable discharge opening has to be provided. Gates and valves are therefore an essential and critical part of many flood control schemes, of reservoir management and the control of water in river courses.

Many types of gate are in successful operation. However few of these may be suitable or economic for a specific situation. The problem is to select and design the most appropriate type and size of gate which will meet the hydraulic, operational, site specific and economic requirements.

Gates and valves control the flow of water. The hydraulic conditions are therefore basic to the success of the installation. This comprises not only the flow under or over a gate but also the upstream and the downstream hydraulics.

Since gates are designed for extreme events, personal experience of their performance under these conditions is limited. Some gate installations have met with serious difficulties in service. The subsequent research and published papers have sometimes been presented at a specialist congress and have not been disseminated to prevent the repeat of flawed design features. This book attempts to present practical experience and available knowledge of gates and valves in civil engineering structures to provide a guide to their selection. Information on details of gate design is presented to help in an assessment of the suitability of a gate for its task. The chapter on valves is intended to assist in identifying the correct type of valve for its duty. Ancillary equipment required in control structures, such as screens, stoplogs and handling equipment are the subject of separate chapters.

The author thanks clients for whom he has carried out projects for permission to use information acquired in the course of his work.

Note on units

Most of the symbols used in the equations of the text have no designated units. Wherever this is the case consistent SI units can be used.

2
Types of gate

In this chapter, gates are divided into two groups, in free surface flow and in conduits. While similar types of gate are found in both groups, the design of gates in submerged outlets, especially at high-heads, is more demanding and requires special consideration. Appendix 2.1 lists the main applications and the advantages and disadvantages of the various types of gates.

While there are many types of gate, a limited number predominate because of their advantages. In open channels, at spillways and barrages, radial gates are the first choice. Except for very large span gates in navigable rivers, new vertical-lift gate installations are infrequent. This does not apply to rehabilitation of old barrages such as the Sukkur and Kotri barrages on the river Indus in Pakistan where old gates are being replaced by similar vertical-lift gates of modern construction.

The bottom-hinged flap gate or tilting gate is sometimes preferred in river courses because it is considered less visually obtrusive than a radial gate. Also its ability to discharge debris over the gate may be important. A radial gate requires a flap section to carry out the same function. In tidal barrages a tilting gate can completely prevent ingress of saline water to the upstream pond or reach. It is often chosen on this account. In undershoot gates under drowned flow conditions a lens of saline water can penetrate upstream against the flow of water.

In conduits, vertical-lift gates are more frequently used than radial gates. This is mainly due to greater flexibility in installation, since radial gates require a large chamber to retract. This is in spite of the hydraulic problems due to gate slots at high velocity flow.

Detail aspects of gate design are dealt with in separate chapters because they apply to a number of different gates. The exception is radial automatic gates where many design considerations are special to this type of gate.

A number of gates have been developed for special conditions, such as tidal barrages or storm surge barrages. So far they have not found more general application.

Gate types which were once common, such as the rolling-weir gate, drum and sector gates, Stoney-roller gates and others not dealt with in this chapter, have been largely superseded because of their complexity, cost and associated civil engineering construction cost.

2.1. Gates in free surface flow

2.1.1. Radial gates

Radial gates are the most frequently used movable water control structures. They consist of a skin plate formed into a segment with a radius about the pivot. The skin plate is stiffened by vertical and horizontal members which act compositely with the skin plate. The skin plate assembly is supported by two or more radial struts which converge downstream to the pivot assemblies which are anchored to the piers and carry the entire thrust of the water load. The resultant of the water load passes through the pivot pins and there is no unbalanced moment to be overcome when hoisting the gate. The hoisting load consists of the weight of the gate, the friction force between the side seals and the seal contact plates embedded in the piers and the moment of the frictional resistance at the pivots.

Radial gates are under most conditions the simplest, most reliable and least expensive type of gate for the passage of large floods. Their advantages are:

- Absence of gate slots. (This benefits pier structural design and hydraulic flow. Pier slots can produce cavitation and at low flows collect silt.)
- Gate thrust is transmitted to two bearings only.
- Less hoisting capacity is required than for a vertical-lift gate.
- Mechanically it is simpler and mechanical equipment usually costs less.
- Location of the bearings protects from damage by debris, simplifies corrosion protection and permits some degree of inspection while in service.
- The superstructure required for the gate lifting gear is generally much lower, and in the case where a roadway spans the sluiceway no additional structure is usually required.
- Stiffer structurally.
- Better appearance.
- No possibility of trash getting jammed in the wheels.

The disadvantages of radial gates are:

- The flume walls are required to extend downstream at a sufficient height to provide attachment for the gate pivot bearings.
- The gate water load is taken by the piers as concentrated loads at the gate anchorages. Integrity of the anchorages and distribution of the load into the piers require special consideration because of this factor. In gates of 150–200 m^2 aspect area and larger, this has led to the use of prestressed steel and concrete anchorages.
- Increased fabrication complexity.

Radial gates will not pass floating material until they are at least 75% open. This can be overcome by adding a flap or overflow section to the top of the gate (Figure 2.1(b)). These are usually operated by an independent motor and hoist gear or separate hydraulic cylinders. The overflow discharge section is curtailed on this type of gate for the nappe to clear the gate arms. Tilting crests may be required when ice floes have to be discharged in spring or where the river carries an abnormal amount of debris during part of the year.

The shape of the tilting crest has to be checked so that flow separation cannot take place in any one position when the crest section is lowered. If the crest section is curtailed the nappe will be vented and problems due to nappe collapse will not occur.

Radial gates have been designed to be submergible to provide overflow (Figure 2.1(c)).

Hydraulic forces acting on a radial gate

In a closed radial gate the resultant hydrostatic forces act through the pivot. This applies to upstream and downstream forces.

Provided the skin plate has been formed in a true radius with the origin at the pivot point there are no unbalanced forces about the pivot due to hydrostatic forces. When the gate is lifted the forces are the mass of the gate with the centre of gravity fairly close to the weir plate assembly, the frictional resistance of the side seals and the frictional resistance of pivot bearings.

The seal friction will be significantly higher on starting compared with the running friction, and this also applies to bushed bearing although not to roller bearing pivots.

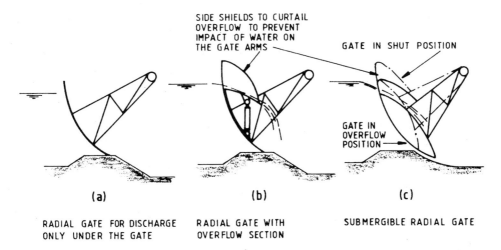

(a) RADIAL GATE FOR DISCHARGE ONLY UNDER THE GATE

(b) RADIAL GATE WITH OVERFLOW SECTION

(c) SUBMERGIBLE RADIAL GATE

Figure 2.1. Types of radial gate

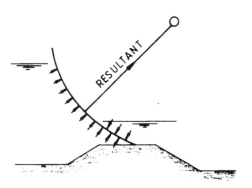

Figure 2.2. Hydrostatic forces acting on a radial gate

Figure 2.4 shows the distribution of pressure head for a radial gate under free discharge conditions. If the pressure curves are available from flow-net analysis or actual measurement on model or prototype, the required components of the forces can be obtained by graphical integration. An approximation may be obtained by assuming hydrostatic distribution of pressure over the gate as indicated by the broken lines of Figure 2.5. The corresponding force components are equal to the enclosed areas multiplied by the specific weight of water and the gate width. This will result in an overestimate of the hydraulic forces acting on the gate. Rouse[1] sets out a method of computing the hydraulic forces by taking into account the discharge characteristics of the gate. This leads to a closer approximation.

Constructional features of gate arms

Gate arms are usually offset (Figure 2.6). This reduces the bending moment on the horizontal girders connecting the arms and, where necessary, permits

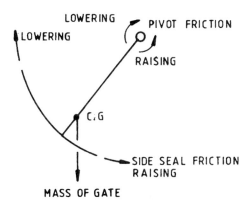

Figure 2.3. Lifting and lowering forces acting on a radial gate

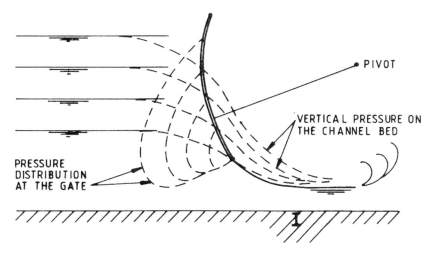

Figure 2.4. Distribution of pressure head for radial gate under free discharge conditions

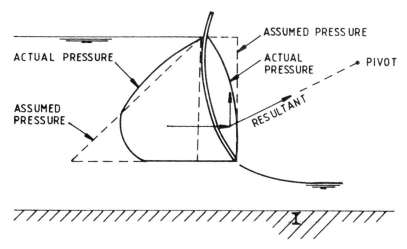

Figure 2.5. Approximation of pressure head for a radial gate under free discharge conditions by assuming hydrostatic distribution of pressure over the gate

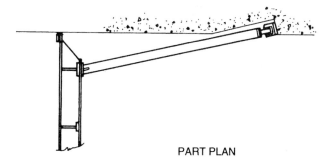

PART PLAN

Figure 2.6. Conventional arrangement of arms of a radial gate

the pivot bearing to be recessed into the pier. In gates operating under drowned-discharge conditions in rivers it also avoids debris being trapped between the gate arms and the piers.

The lowest gate arm is usually positioned as low as possible for structural reasons (Figure 2.7). Under drowned-discharge conditions severe turbulence is set up in the stilling basin and an unsteady roller occurs[2]. If the roller acts on the submerged gate arms or on other structural members vibration is likely to occur[3]. In this case hydraulic considerations should override structural priorities so that members are disposed in the most efficient manner (Figure 2.8).

The hydrodynamic effect on the gate is usually ignored in calculations of structural strength as well as the hydrodynamic downpull forces which become more significant when radial gates are used as culvert valves under high-head conditions (Figure 2.9).

Conventional calculations of the strength of the weir plate assembly ignore the additional strength and rigidity due to its curvature. To take this into account in larger gates involves a degree of complexity of analysis which most designers do not consider warranted. The steps in the analysis of a weir plate assembly of a radial gate are:

- Panel stress due to the hydrostatic load (1).
- The design of the intermediate horizontal stiffener beams (2) where each of these members is assumed to transfer one quarter of the load of its two associated panels to the vertical stringers (3).
- Load imposed on the main horizontal beams (4) which tie together the gate arms (5).
- The horizontal stiffener beams (2), the vertical stringers (3) and the horizontal beams (4) are all designed to act compositely with the skin plate.

The method of calculating panel and combined stresses is set out in Chapter 5.

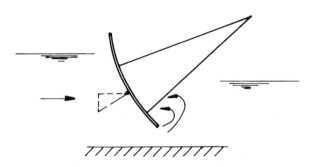

Figure 2.7. Positioning of arms of a radial gate

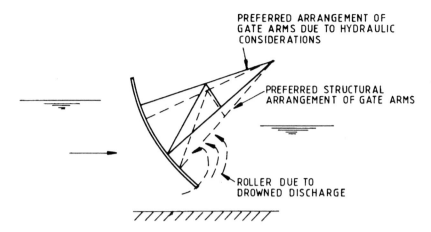

Figure 2.8. *Effect of reverse roller on the lowest arm of a radial gate under drowned discharge conditions*

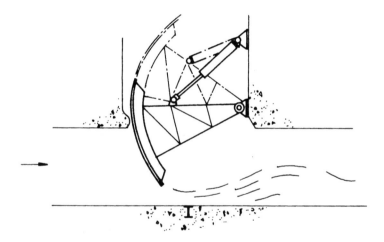

Figure 2.9. *Radial gate used as a culvert valve*

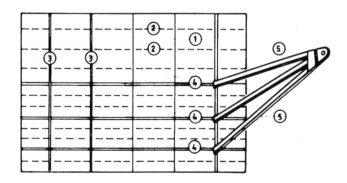

Figure 2.10. *Steps in the analysis of a weir plate assembly of a radial gate*

Radial gates can be operated by electric motor driven hoists or by hydraulic rams. Some arrangements of hoists and of hydraulic rams are described and illustrated in Chapter 6.

2.1.2. Radial automatic gates

Radial automatic gates have been built and have operated during the last 50 years[4]. They require no outside source of power, are simple and reliable. Provided that they are correctly designed hydraulically they will operate consistently with minimum attention and require very little maintenance. In general, radial automatic gates can be installed where upstream level control is required under variable inflow conditions and where the downstream stage does not rise disproportionately quickly[5]. They can also be used to control downstream level in irrigation canals.

Water level control

The gates are actuated by changes in water level and either upstream or downstream water level can be controlled. The former is usual in rivers and the latter in irrigation channels. Figures 2.11 and 2.12 illustrate the most frequent arrangement of upstream water level control. The radial arms which support the skin plate assembly extend downstream of the pivots to carry counterweights which balance the gate. Operation is through displacers in each side pier. The displacers are attached to the radial arms by pivots, so that the forces exerted on the gate, due to changes in displacer submergence, cause the gate to open and close. The displacers have to overcome side seal and pivot friction which act in both directions as shown in Figure 2.3.

The gate control system comprises an intake upstream of the gate protected by a screen. The intake is controlled by a sluice valve and discharges into the weir chamber. The discharge over the weir flows into the displacer chambers which are interconnected by a pipe passing under the sluiceway. An outlet from the displacer chamber, also controlled by a sluice valve, returns the flow downstream of the gate. This is shown in Figure 2.12.

An increase in the upstream level causes flow through the intake pipe into the weir chamber and discharge over the weir. This in turn causes a rise in level in the displacer chambers and flow to the river downstream of the gate. Due to increased buoyancy of the displacers, the gate rises and discharge occurs under the gate. Rise of the gate may cause a slight drop in upstream water level, which will result in reduced flow over the weir and hence a lowering of the level in the displacer chambers. The gate may close slightly as a consequence until balanced conditions are achieved. The stability of the system is provided by head loss in the inlet system and side seal friction.

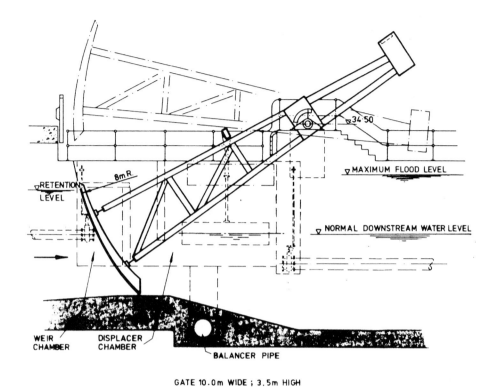

Figure 2.11. *Radial automatic gate for upstream water level control*

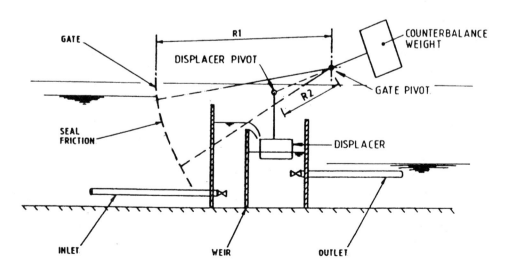

Figure 2.12. *Radial automatic gate control system*

The inlet system

The inlet is arranged some distance upstream of the gate. It should be positioned facing downstream and at two-thirds of the retention level to minimise obstruction of the inlet due to flotsam. The pipe is usually made large to enable rodding out. The head loss which should be 50 to 65 mm is provided by the setting of the inlet valve. Figure 2.13 shows a typical arrangement.

The control weir

The control weir acts as an amplifier and takes the form of an adjustable sharp edged weir set 50 to 65 mm below the retention level. A convenient way to achieve a long weir is shown in Figure 2.13. A stub wall at the end of the weir permits aeration of the nappe.

Displacers and displacer chambers

The flow from the weir is baffled to provide quiescent conditions in the displacer chambers. Movement of the gate lip relative to that of the displacers is amplified by the ratio of the gate radius to the radius of suspension of the

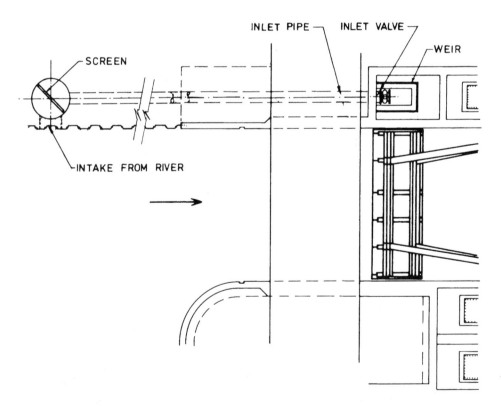

Figure 2.13. Inlet system for the control of a radial automatic gate

displacers. Similarly the frictional resistance of the side seals referred to the displacer buoyancy is amplified by the same ratio. Since the direction of the friction force changes on opening and on closing of the gate, the change in water level within the displacer chamber required to reverse the movement of the gate will be double.

Upstream level retention within close limits is possible which can be as low as 12 mm, although under high flow conditions 50 to 80 mm rise will be required for the gate to discharge the increased flow.

Under flood conditions when the upstream level cannot be retained (a condition which is also associated with a substantial rise in downstream level), the inlet weir is drowned and the gate operates according to the ratio R_1/R_2 as shown in Figure 2.12. The gate will therefore lift more rapidly than the rise in flood water and will cut out of the water to permit unobstructed flow under these conditions.

If the initial difference between upstream and downstream level is 2 m or less, the stage–discharge relationship must be known when the design is carried out, so that the gate control system is arranged to avoid the down-stream level taking over control of the gate during some stage of the rising flood. This can also be of importance during a falling flood when the downstream level falls more slowly than the upstream level. This can result in a condition when the downstream water level keeps the gate open, causing loss of retention level.

The displacers are designed so that their specific gravity is 1.05 to 1.1. The construction takes the form of a stiffened box with watertight access covers for loading weights. The total assembly forward of the pivot is out of balance to the extent that the displacers are half immersed when the gate is in a steady condition. Equal forces are then available for opening and closing the gate. With this arrangement the mass of one displacer must at least be equal to the friction force of one side seal $\times R_1/R_2$ plus the pivot friction. A gate with displacers sized on this basis would have no margin and would provide very coarse level control. With displacers designed three times this size they will give good service.

The wide piers required to accommodate displacer chambers can be a disadvantage in a sluice installation. If self-aligning roller bearings are used for the pivots the effort at the displacers to overcome friction is negligible. For bronze bushed bearings it will amount to 1–2.5 kN at the displacer. It is the usual practice to use displacers. There are some gates where floats have been used. Figure 2.14 shows a design where floats are maintained stable by the link 'a' so that pivots 'b' and 'c' and the link form a parallel motion. In a 6 m wide gate, as shown in Figure 2.11, using floats would have resulted in an overall saving of 2.8 tonne.

To ensure consistent operation throughout the range of travel of the gate, the centroid of the counterweight, the centre of the pivot bearing, the centre line of the displacer suspension pivot and the centroid of the skin plate

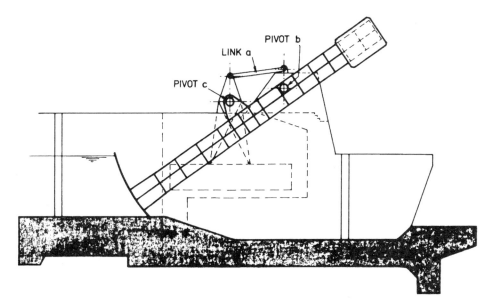

Figure 2.14. Radial gate using floats (11.5 m radius, 15 m width)

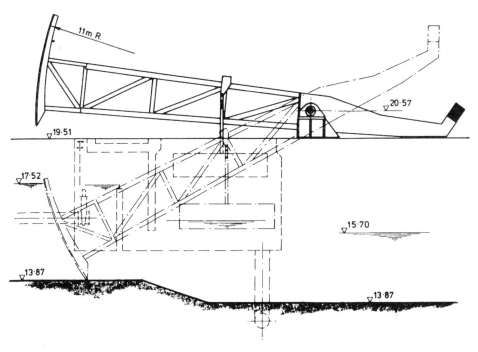

Figure 2.15. 15.24 m wide gate, Pulteney Sluices, City of Bath Flood Protection Scheme

assembly should be in one line (Thorne[4]). Departure from this is possible, provided the gate operation is calculated separately for different openings.

Pivots

Most gate manufacturers use phosphor-bronze bushed bearings. These have a coefficient of friction of 0.1 to 0.12 starting, reducing to 0.07 during rotation. The gates in Figures 2.11 and 2.15 incorporate self-aligning roller bearings. Their coefficient of friction is 0.0018. Using self-aligning roller bearings eliminates uneven bearing pressure in a bushed bearing due to deflection of the pivot pin. Self lubricated bearings, mentioned in Chapter 7, and illustrated in Figure 7.15, significantly reduce maintenance. They are available as plain bushes or self-aligning. Their coefficient of friction is similar to that of phosphor-bronze bearings, although limited site tests suggest that values are about 30% below manufacturers' published data.

Counterbalance

This can take the form of a reinforced concrete beam with pockets cast into the upper section so that final balancing of the gate can be carried out. Where environmental conditions require a less obtrusive appearance, as at Pulteney Weir in the City of Bath (Figure 2.15), the kentledge can be of cast iron sections bolted to a structural beam.

Downstream water level controlled gates

Gates for downstream water level control use a float chamber downstream of the pivot (Figure 2.16). These are suitable only for small gates, because the discharge under the gate causes turbulent conditions. Larger gates have to be severely damped, or displacer chambers in the piers have to be constructed using an intake positioned downstream of any turbulence. It is usual to baffle the intake. Gates of this type are not subject to size limitations.

Causes of gate instability and malfunction

Gate instability can be due to:

- Insufficient head loss in the inlet system.
- Insufficient side seal friction.
- Limited ponded up water.
- Reflux of flow around piers dividing two sluiceways.

  Malfunction of gates can be caused by:

- Blocked inlet system or screen.
- Lodgement of an obstruction at the sill beam.
- Flooded displacer.

Dividing piers between sluiceways

If the length of dividing piers between sluiceways is short, the discharge from one sluiceway may, under drowned-discharge conditions, reflux around the pier and set up oscillating waves acting on transverse stiffeners of the gate in the adjoining sluiceway. This can set up harmonic motion of the second gate which, in turn, can affect the first gate which will respond in a similar manner, but out of phase. Such motion will not amplify more than 150 to 300 mm. It can be a nuisance to river navigation. In an installation in the River Lee in Essex, the dividing pier had to be extended in order to eliminate this effect so that its length downstream of the gates was increased from 15 m to 23 m.

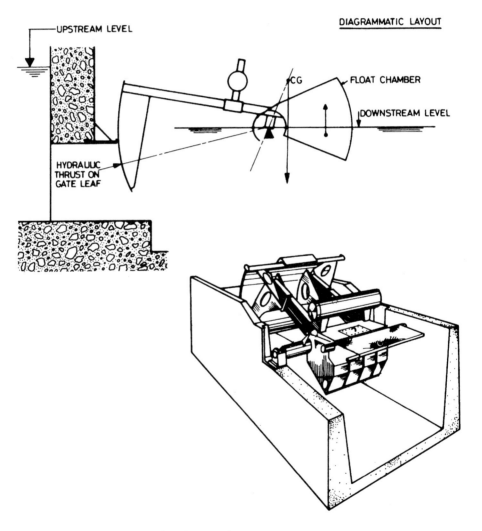

Figure 2.16. Downstream level control gate

Computer programme for radial gates

Because unstable operating conditions can arise in radial automatic gates, and because design and testing for instability can be carried out only on a trial and error basis, varying parameters in turn, a computer programme forms a useful design tool so that changes in variables can be rapidly examined.

The programme output has to determine the relationship between upstream and downstream water levels, gate opening and the discharge under the gate. The relationships have to be computed for the rising and falling river or reservoir stages and also when the gate is clear of the water.

Such a programme was run for the gate shown in Figure 2.11 while the gate was in the course of preliminary design. The computer run demonstrated that the gate cut out of the water too quickly because the rapid rise in downstream level took over the control of the gate. This caused an unstable condition because the upstream level then dropped below the retention level. The programme was then re-run increasing the water level of the displacer chamber and the outlet from the displacer until stable conditions were obtained.

2.1.3. Vertical-lift gates

The advantages of vertical-lift gates are:

- A short length of flume walls.
- Distribution of the gate water load.

The disadvantages are:

- The requirement for gate slots.
- Possibility of trash getting jammed in the wheels.
- Overhead structure.
- Wheels which have to rotate under water.

The majority of vertical-lift gates are counterbalanced to reduce the hoisting load. To prevent the counterbalance from entering the water when the gate is lifted, the counterbalance is reeved 2:1 so that it travels for only half the distance. This results in an additional load on the superstructure of the order of 2.7 times the mass of the gate, and requires a substantial support structure, adding to the cost of the gate installation.

Most gates of this type used in open channels are of the fixed roller type (Figure 2.17). In gates in open channels, rollers are usually spaced out to take an equal load of the hydrostatic forces acting on the gate. Roller alignment is critical, uneven contact of a roller can overload adjoining rollers. One method of adjusting rollers to ensure equal contact on the roller path is shown in Figure 7.12.

Downstream sealing of a gate is preferred because the water load compresses the seals. Upstream sealing is required where a gate is located

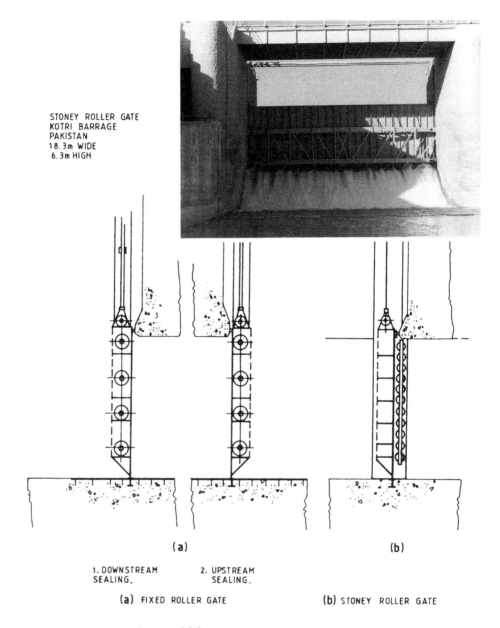

STONEY ROLLER GATE
KOTRI BARRAGE
PAKISTAN
18.3m WIDE
6.3m HIGH

(a)   (b)

1. DOWNSTREAM   2. UPSTREAM
   SEALING.   SEALING.

(a) FIXED ROLLER GATE   (b) STONEY ROLLER GATE

Figure 2.17. Types of vertical lift gate

upstream in a shaft, and access for inspection of the tunnel and the gate is via the shaft.

Figure 2.17(b) is a diagram of a Stoney-roller gate. The roller trains, one on either side within the gate slots, move at half the speed of the gate and are reeved 2:1 on their suspension. In practice, the load distribution from the gate via the rollers to the track is not even, and the permissible contact

(Hertz) pressures for Stoney rollers is half of the pressure for load roller wheels (see also Section 7.2). The rolling load is transmitted from the rolling face on the gate directly to the track face and Stoney roller axles should be subject to only nominal rotational friction. This is not the case, since the rollers deform elastically imposing a load on the axles or bushes, if they are fitted. In addition, inaccuracies in the alignment of the rollers can impose considerable additional loads. Breakdown of individual rollers has occurred as a consequence. The slack in the tracking of a Stoney-roller gate can cause gate vibration, especially under conditions of high velocity flow. Stoney-roller gates have fallen into disfavour. There are however many gate installations of this type which are being refurbished or will require replacement.

Vertical-lift gates can also be fitted with overflow sections (Figure 2.18) where limited overflow is required. Where substantial overflow must be effected, a Hook-type gate is used (Figure 2.19). This results in considerable complexity of construction and high accuracy of manufacture is required to maintain the seal between the two sections.

Hook-type gates use either a single hoist, so that the upper section is hoisted first and when it reaches the full extent of travel it moves together with the lower section, or two hoists. The latter arrangement is required for combined over and underflow when the gate is to be used for underflow without moving the upper leaf. This may be required for aesthetic reasons. The hydraulic conditions caused by combined over and underflow can induce gate vibration and the range of operation of such a gate may have to be restricted.

For long-span vertical-lift gates, where the skin plate structure is backed by girders and where there is drowned-discharge, similar considerations to those mentioned earlier in this chapter for arms for radial gates apply. The turbulent discharge conditions downstream of the gate and an unsteady roller, if it develops, can act on the structural members and cause local or general gate vibrations. This is discussed further in Chapter 10.

Larger vertical-lift gates are often manufactured in sections with an articulated joint between sections. Figure 2.20 shows how this can be effected and also illustrates the sealing between sections. Apart from the saving in site welding that such designs permit, it is often possible to mount four wheels on any one section and omit adjustment of the rollers.

An articulated vertical-lift gate has to make provision for transfer of shearing forces from one section to the other so that racking forces do not shear the seal between sections. Transverse guidance of vertical-lift gates is provided by separate guide rollers or slides.

2.1.4. Rolling-weir gates

Figure 2.21 illustrates a rolling-weir gate where 'a' is the drum and 'b' a lip to effect flow separation. The gate has a spur gear segment 'c' which engages

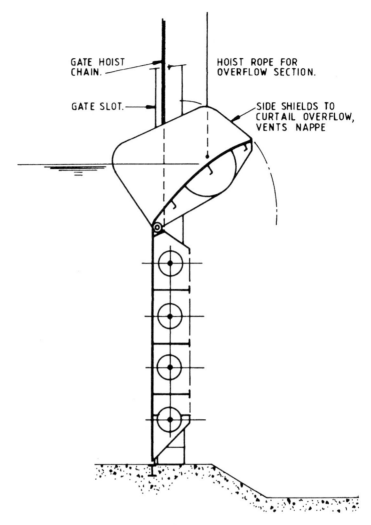

Figure 2.18. Vertical lift gate with overflow section

with a rack 'd'. To raise or lower the gate it is lifted by chains 'e' which make it roll upwards on the rack 'd'.

Rolling-weir gates were designed for wide sluiceways where their structural rigidity and high torsional resistance were advantageous. Some gates have been designed so as to be submergible to clear ice floes in Spring. This can impose difficulties in flow control at low discharge.

The gate is complex to manufacture, imposes difficulties in designing effective side seals and is vulnerable to jamming if debris becomes lodged on the rack. Few, if any, drum gates have been manufactured during the last twenty years on this account.

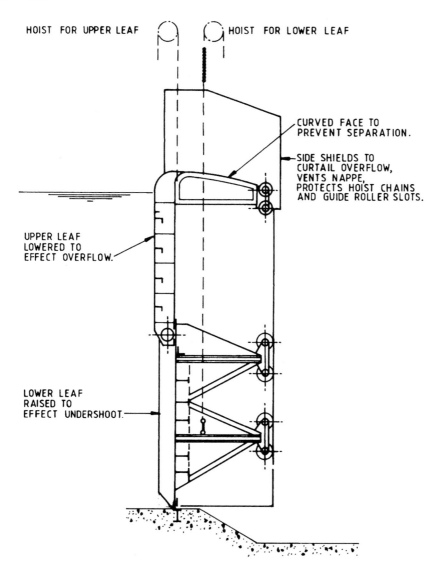

HOIST FOR UPPER LEAF   HOIST FOR LOWER LEAF

CURVED FACE TO
PREVENT SEPARATION.

SIDE SHIELDS TO
CURTAIL OVERFLOW,
VENTS NAPPE,
PROTECTS HOIST CHAINS
AND GUIDE ROLLER SLOTS.

UPPER LEAF
LOWERED TO
EFFECT OVERFLOW.

LOWER LEAF
RAISED TO
EFFECT UNDERSHOOT.

Figure 2.19. Hook type gate

2.1.5. Overflow gates

Flap or tilting gates

Bottom-hinged flap gates are used in tidal rivers to prevent the ingress of saline waters or where special environmental considerations apply[6]. They are sometimes selected for reservoir spillways on the consideration that they can be made to operate in an emergency on a 'fail safe' basis. This is effected for gates operated by hydraulic cylinders so that the oil in the annulus side of the piston is passed to the cylinder side by opening a bypass valve. Because

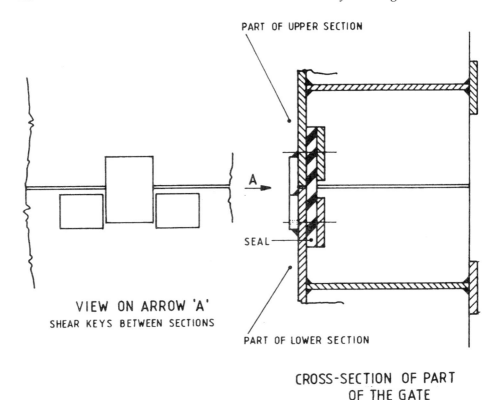

PART OF UPPER SECTION

A

SEAL

PART OF LOWER SECTION

VIEW ON ARROW 'A'
SHEAR KEYS BETWEEN SECTIONS

**CROSS-SECTION OF PART
OF THE GATE**

Figure 2.20. Vertical lift gate manufactured in sections

the volume of oil in the annulus side is less than that in the cylinder side for a given length of stroke, the cylinder may have to be vented to atmosphere or a degree of cavitation in the cylinder may have to be accepted.

For fish-belly flap gates in river courses or barrages, where the downstream water level can rise above the level of the gate hinges, complete 'fail safe' opening is possible only if water is admitted to the fish-belly section, otherwise the gate will open only to the stage when the fish-belly becomes buoyant, that is when the mass of water above the gate and its submerged mass are equal to the buoyancy and pressure under the gate. It is not usual to admit water into the fish-belly section because it is difficult to repaint, and silt and sediment can accumulate in the section. Because tilting gates discharge by overflow, floating debris can be cleared without having to introduce a separate flap or tilting section as in bottom-discharge gates.

A disadvantage of bottom-hinged flap gates in river courses or tidal barrages is permanent immersion of the hinge bearings and inability to inspect the bearings and the hinge seal where the downstream water level is always above the level of the hinges, except by placing stoplogs. Figure 2.22 illustrates different versions of bottom-hinged flap gates.

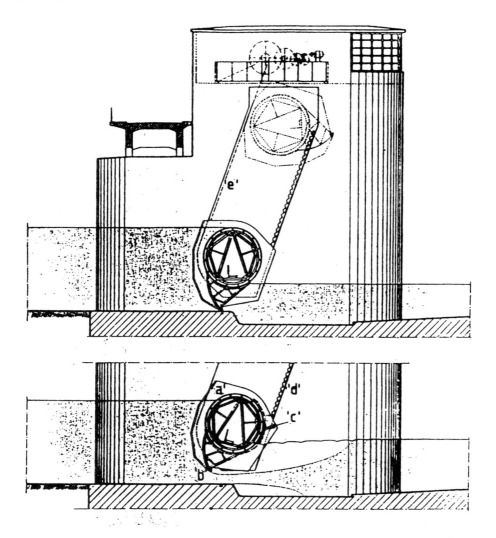

Figure 2.21. Principle of rolling-weir gate

Figure 2.22 (a) shows a flap gate operated by a hydraulic cylinder positioned underneath the gate. This arrangement can present difficulties in servicing the cylinder where the downstream water level does not fall below the downstream bed level. Figure 2.22 (b) shows an arrangement where the operating cylinder is housed in the piers. A torsion tube transmits the operating force laterally. Where the operating shaft passes into the pier chamber the seal adds to the complexity of this design. Figure 2.22 (c) shows a single operating cylinder positioned at the side of the gate. The structure of such gates is made torsionally rigid and often takes the form of a 'fish-belly'.

Bottom-hinged gates open by tilting in a downstream direction until they lie flat or form the required crest profile. They are usually sealed along the

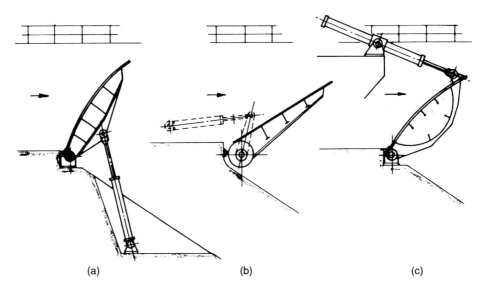

(a)   (b)   (c)

Figure 2.22. Different versions of bottom-hinged flap gates operated by hydraulic cylinders

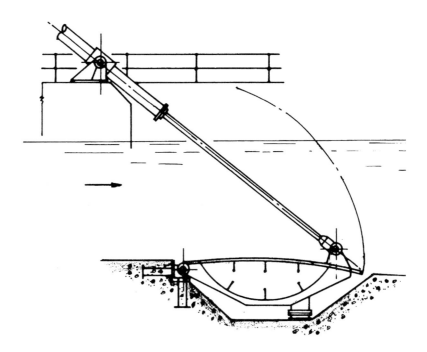

Figure 2.23. Recess in sluiceway to permit complete retraction of flap gate under flood conditions

hinged edge by a flexible flat rubber strip clamped to the embedded sill member and extended to rest against the upstream face of the gate skin plate. Because the gap to be sealed varies throughout the movement of the gate, it is possible to extrude the seal under maximum hydrostatic head, and designs clamping the seal both to the sill and the gate face plate have been evolved. Side sealing over the full travel of the gate is essential, since debris in the water tends to be drawn into any gap. If the gates are made to close against an abutment in the sluiceway in order to seal by making direct contact with the upstream face of the recess, any material trapped is compressed or wedged and there is a danger of the gate becoming jammed. Recessed abutments are not recommended for hydraulic reasons. It is therefore good practice for side seals to sweep the sluice walls throughout the total movement of the gates. This requires machined side plates, of either cast or fabricated construction, embedded flush with the civil structure.

Bottom-hinged flap gates which seal against the concrete face of an abutment or a pier have been constructed. This requires that the tolerance of the concrete face is ±3 mm along both axes over a distance of 2 m. While this saves the cost of embedded side plates, it increases the wear of the seals and causes greater construction costs to attain the required accuracy and finish of the concrete. Seal designs for bottom-hinged flap gates are illustrated in Chapter 7 (Figure 7.8).

Where gates have to be fully retracted the shape of the skin plate has to be formed so as to avoid flow separation and sub-atmospheric pressure. This is a potential trap for debris. Where operational reliability is paramount, it may be necessary to provide means of clearing debris without dewatering the sluiceway or using a diver. This can be done by jets of water, by compressed air or by flushing with river water. No published information regarding practical experience in the use of compressed air distribution systems for this application is available. If river water is used to flush out the recess, the design has to avoid the creation of eddies and dead pockets which will cause the transfer of debris from one section in the recess to another without dislodging it. Debris will be trapped behind the overflow jet and a bypass system, as shown in Figure 2.24, is necessary to clear flotsam. The gate is elevated to stop overflow when flushing out.

Tilting gates can be manually or motor operated through screws or actuated by means of two hydraulic cylinders on either side of the gate or by a single cylinder centrally positioned or only on one side. Small gates are more frequently operated by ropes. The arrangement of the hoist machinery is then similar to that shown in Figure 6.1.

Rising screw type gearing, with twin lifting screws operated from a central headstock, gives the required large mechanical advantage and also provides a self-locking feature which resists the water load tending to reverse drive the gears. The use of exposed lifting screws, either in the sluiceway or in a

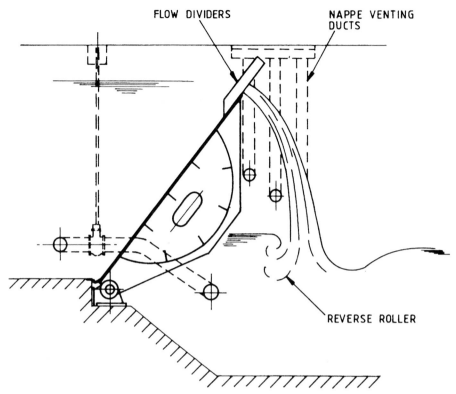

FLOW DIVIDERS   NAPPE VENTING DUCTS

REVERSE ROLLER

Figure 2.24. Bypass system for flushing debris accumulated downstream of a flap gate

recess in the sluiceway walls, is a potential source of malfunction. Totally enclosed screws operating in an oilbath are preferred.

Oil hydraulic operation by placing the ram under the gate makes it possible to dispense with an overhead structure. Maintenance of the ram then requires dewatering of the sluiceway. This is often considered a major drawback of this configuration.

At the Bala sluices of the Dinorwic Pumped Storage Scheme, the planning requirements stipulated that no overhead structure should be provided. The gates were therefore designed as a torque tube structure, with the pivot shaft extended into the hollow piers where they were operated by a hydraulic cylinder acting on a lever.

A special case of a bottom-hinged flap gate is the velocity-control structure in the river Orwell at Ipswich[7]. In its elevated position, the gate acts as a submerged weir reducing the cross-sectional area of the river, reducing the flow and scouring action on the river bed.

Automatic tilting gates have been constructed with counterbalance above or below the gate. The counterweight is arranged to balance the overturning moment of the upstream water load at normal retention level. With a rise in

Figure 2.25. Bala sluices of the Dinorwic Pumped Storage Scheme

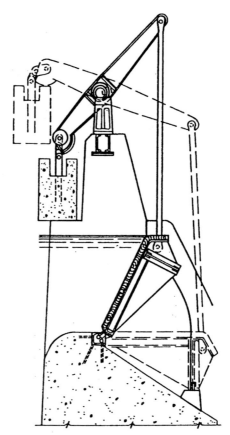

Figure 2.26. Automatic tilting gate

level the gate becomes overtopped and the overturning moment is increased. When this overcomes the resistance of the counterweight, the gate opens. By careful proportioning and positioning of the counterweight system and the pivot point, these gates can be arranged to open and close in a series of movements on a rising and falling upstream level. The degree of control is not accurate and a significant variation in level is required to effect full gate travel. If on discharge the downstream level starts to rise, thus creating an overturning moment tending to close the gate, it may be impossible to prevent the upstream water level from rising.

The gate arrangement shown is inherently hydraulically unstable. Disturbance may be set up due to wave motion in a reservoir; or a pulsating surge or wave may be set up in the upstream reach of a long approach channel of uniform section. Surging can be set up due to level drawdown immediately upstream of the gate following a downward movement. The loss of water load on the gate then causes a closing movement, and if the frequency of gate oscillation coincides with that of the surge wave the gate movement is accentuated and can become dangerous.

The only damping force present is the friction of the side seals, hence hydraulic dashpots have to be added to the counterweight system. A control system which is more stable utilises the same principle as radial automatic gates by arranging the counterbalance weights so that they act as displacers.

The nappe has to be vented to prevent gate vibration and nappe collapse. Flow dividers are used to vent the nappe under moderate overflow conditions. The design and spacing of flow dividers is critical[3]. The dividers must project beyond the gate lip and they must be wide enough to form an adequate opening in the nappe for the admission of air. The flow of water over the gate lip expands as soon as it is no longer in contact with the divider and tends to close up.

Experimental work and prototype trials have been carried out to study nappe oscillation and resulting vibrations[8-13]. Model studies to Froude scale of bottom-hinged flap gates incorporating flow dividers are ineffective in preventing self-excited nappe oscillations and resulting vibrations[14]. The spacing of flow dividers is important. Pulpitel[15] gives some information on flow dividers which can be applied in practice. The initial flow dividers were spaced at 2100 mm centres. Additional ones were introduced between the original flow dividers. They projected 280 mm beyond the skin plate and had a crest width of 300 mm and a width over the tapering side sections of 450 mm. The maximum overflow depth was 720 mm.

When the head of water above the gate lip is appreciable, flow dividers become ineffective[16] and additional venting through the sluiceway walls or the piers has to be provided. A method of calculating the air demand of an overflow jet is given in Chapter 9.

Top-hinged flap gates

Top-hinged flap gates are used in tidal structures to prevent flooding of an inland region by sea waters during rising tides or flood surges and to permit inland waters to drain off into the sea during ebb tide. They are also used in culverts and pumped drainage outfalls to rivers.

They do not require an outside source of power and operate automatically. The construction of the gates is simple and little maintenance is required. The gates will not entirely exclude ingress of saline water if the downstream water level rises above the sill during discharge under the gate, when a lens of saline water can penetrate upstream against the flow.

They control water in one direction only and perform like a non-return valve. They cannot control upstream level. In stormwater discharge this facility is not required. Top-hinged flap gates can be operated under clear discharge conditions or drowned. When designing gates of this type a gravity bias is required in the closed position so that the gates close immediately before reversal of flow occurs. This can be effected by slanting the closed gate position, an arrangement which was investigated as part of the Severn Tidal

Project Study (Bondi scheme) or the flap may be in the vertical position when closed, with the bias to closure due to eccentric hinges, Figure 2.27.

Flap gates using an elastomeric gate leaf (Figure 2.28) deflect to provide the fluid passage for the flow of storm water. Under flow conditions top-hinged flap gates deflect the discharge downwards and where the gate sill is close to a river bed or the bed of an estuary, scour can occur. The arrangement of a slanted gate provides part of the anti-scour apron and can offer some economies in civil engineering construction. The apparent gain can be cancelled out where there is a requirement for stoplog grooves downstream of the gate.

In a vertical-leaf arrangement the closure bias can be provided by an eccentric hinge, or by a weight mounted ahead (downstream) of the flap. If the discharge conduit runs full, the eccentric hinge opens a gap which will cause flow over the gate leaf in the open position. However, the eccentric hinge arrangement reduces the total mass of the flap compared with the weighted flap and therefore provides an increased discharge for a given opening. This is due to the relationship between discharge and mass of the flap. At a given gate opening these are exponentially related.

Sealing is usually effected by face-to-face contact between the flap and body of the gate. This requires exact positioning of the hinges. To ensure even contact between the sealing faces, the pivot lugs are made adjustable. One embodiment of this is shown in Figure 2.29. The cushioning of the gate leaf in Figure 2.32 is effected by the movement of the projecting section of the flap in the seat. This acts like a piston moving in a cylinder. The extent of cushioning is determined by the clearance between the faces. The section in the seat in contact with the flap must be tapered because the flap moves in an arc. If elastomeric seals are fitted they should not be located on the face of the flap because this causes the flow to separate. Seals should therefore be mounted on the frame. Under flow conditions the flap rides the discharge jet. Under free discharge, as well as some conditions of drowned-discharge, a reverse roller forms at the lip of the flap (Figure 2.30). This is similar to the discharge conditions at underflow gates of the radial- and vertical-lift types. If the conduit flow is near full or under supercharged conditions, transverse flap stiffeners at the lip of the flap can cause flow reattachment which is likely to cause vibration[6]. Transverse flap stiffeners should be set back and the lip of the flap should be stiffened by webs leading from the transverse stiffener to the leading edge.

When a series of flap gates are close to one another, such as in estuarial tidal outfalls, the sideways discharge under the gate causes flow interference between adjoining gates and results in hydraulic losses (Figure 2.31). If the discharge through the gates has to be maximised, the losses have to be reduced by training walls.These should extend as far as the length of the sector swept by the opening of the flap.

The theoretical treatment of the stage–discharge relationship of a flap

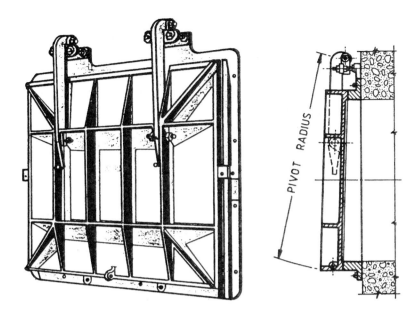

Figure 2.27. Top-hinged flap gate

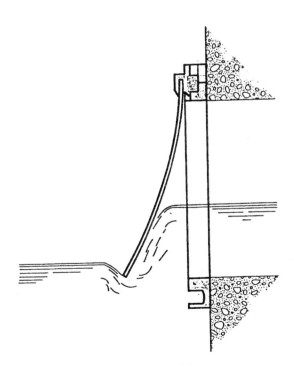

Figure 2.28. Top-hinged flap gate with elastomeric leaf

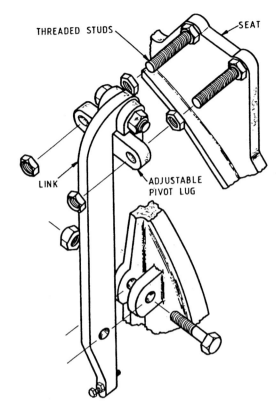

Figure 2.29. Adjustable pivot lugs for a top hinged flap gate

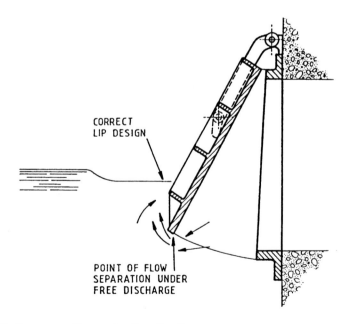

Figure 2.30. Reverse roller at the lip of a flap gate

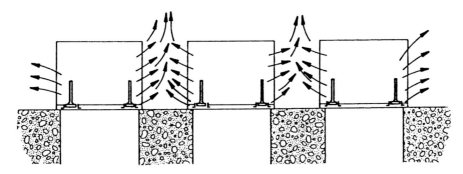

Figure 2.31. Flow interference due to sideways discharge between adjoining flap gates

gate by Pethick and Harrison[22] presupposes that there is no sideways discharge. It is reproduced in Chapter 9.

Wave action or flow reversal will cause rapid movement of the flap and can result in severe slamming. This can be damped by a hydraulic cushion, Figure 2.32, or by an oil hydraulic damper. For a large gate it can be of the oleo-pneumatic kind, or a torsional damper can be introduced in the hinge assembly.

Another method of suspending top-hinged flap gates was used on the Ishmalia gates. This incorporated a cycloidal rocker-type bearing, instead of pivots or hinges, and proved reliable in service.

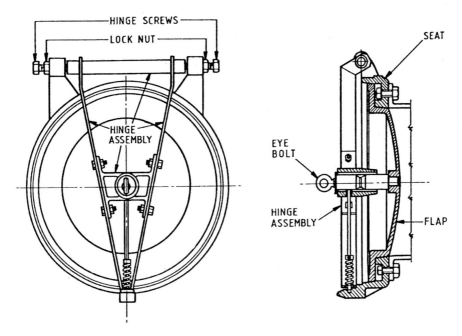

Figure 2.32. Top hinged flap gate hydraulically cushioned

2.1.6. Barrage gates

Bottom-hinged buoyant gates (Venice barrier type)

The design of the Venice Barrier Gates was evolved to avoid any piers within the navigation ways from the Adriatic Sea into the Venice Lagoon. Cruise ships up to 30,000 tonne displacement pass through the Lido passage to the port of Venice and very large supertankers enter the Lagoon through the Malamocco opening.

The gates recess into caissons in the navigation ways (Figure 2.33). To raise them, compressed air is admitted into the gates and water is expelled causing the gates to rise into their operating position at an angle of 50° to the horizontal. The gates can withstand a differential head between the waters of the Adriatic and the Lagoon of up to 1.8 m. This is due to their buoyancy and mass. The four barriers, each 400 m wide, are comprised of gates 20 m wide, moving independently of one another. The gates and the barriers (Figure 2.34) have been described by Lewin and Scotti[17].

Previous schemes of barrages where gates rested in caisson in the navigation way were considered vulnerable to bed material and sediment accumulating in the caissons. In the case of the Venice barriers this was solved by a hydraulic sediment ejector system[18].

Rising-sector gates (Thames barrier type)

A novel concept of a rising-sector gate was developed for the Thames Barrier at Woolwich. Figure 2.35 shows the principle of gate operation. Figure 2.36 shows diagrammatically the gate operating mechanism and a section along the centre line of one of the navigation span gates.

The barrier is formed by four 61 m and two 31.5 m wide rising-sector gates and four 31.5 m wide radial gates. The four 61 m wide gates protect the navigation passages. The barrier and the gates were the subject of a conference and a seminar[19,20]. The papers presented at the conference and the seminar give detailed information of the gates, their operating machinery and the barrage services.

Large span vertical-lift gates

A large span vertical-lift gate forms the River Hull Tidal Surge Barrier (Figure 2.37). When the gate is hoisted to its uppermost position it rotates and comes to rest in the horizontal plane. This reduces the overall height of the support structure. This method of storing a gate in the fully open position is sometimes adopted for vertical-lift lock gates. The upstream lock gate of the Kotri Barrage on the river Indus in Pakistan is an example of this type of construction.

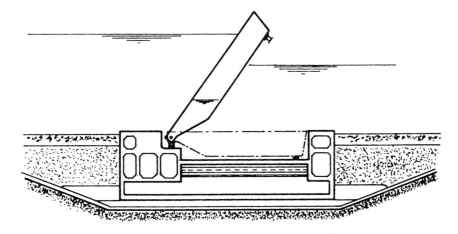

Figure 2.33. Venice Barrage buoyant gate

Lock gates acting as storm surge barriers

Two main types of lock gates are used to protect river mouths from high tides or storm surges. These are mitre gates and vertically-hinged sector lock gates. A few examples of small caisson gates and pointing gates are also used for this purpose.

At the storm surge barrier of the river Hunte in North Germany, conventional mitre gates are arranged in two pairs in parallel. The navigation passage is 26 m wide protected by the 12.7 m high gates. Two radial gates of 20 m span are disposed in parallel on each side of the navigation opening. In all cases the second gate or pair of gates acts as a backup to the seaward gate or pair of gates.

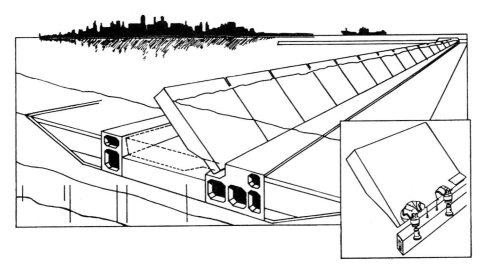

Figure 2.34. Barrage across navigation opening into Venice lagoon

The disadvantage of mitre gates when used for flood protection is the requirement for near balanced conditions for opening and shutting. This makes it impossible to operate the gates in anticipation of a flood. It is presumed to be one of the reasons why two pairs of mitre gates form the

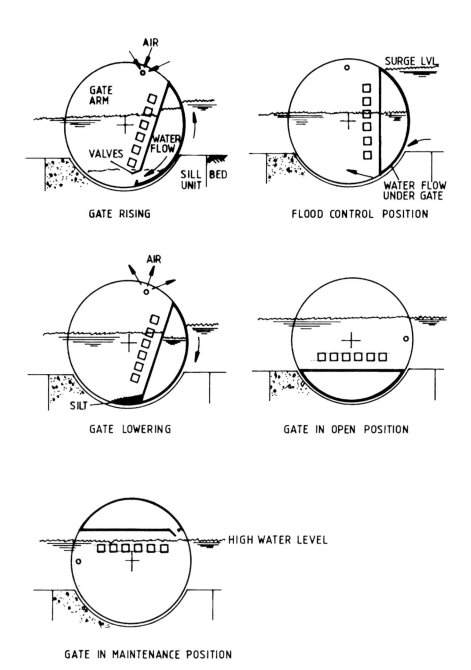

Figure 2.35. Principle of operation of the Thames Barrier gates (after Ayres[19])

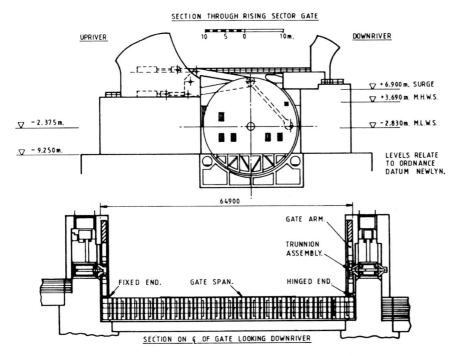

Figure 2.36. Thames Barrier gate (after Clark and Tappin[20])

main part of the Hunte barrier. In the event of failure of one pair of gates to close, there is little time to activate emergency procedures. Another disadvantage is the heavy mitre thrust exerted by large gates, which is considerably in excess of the hydrostatic load on the gate. This affects the cost of the civil engineering works.

These disadvantages are not present in the vertically-hinged sector lock gate. Figure 2.38 shows the gates for a 24 m wide 12 m deep navigation passage for storm surge protection of Tokyo City.

In sector lock gates the skin plate is formed in a true radius and the resultant forces acting on the gate pass through the hinges and produce no unbalanced moment tending to open or close the gate. The river level can be higher than the estuary level or the reverse can occur without affecting the sealing of the gate. Opening and shutting the gate against flow in either direction can be carried out. A disadvantage of this type of lock gate is the large recess required in the river wall to accommodate the gate leaves when they are opened. Very large sector lock gates safeguard a river on Rhode Island on the Atlantic coast of the USA against surges due to hurricanes.

2.1.7. Caisson or sliding gates

Caisson gates move horizontally on rollers and are retracted when not in operation into rectangular chambers at right angles to the navigation way.

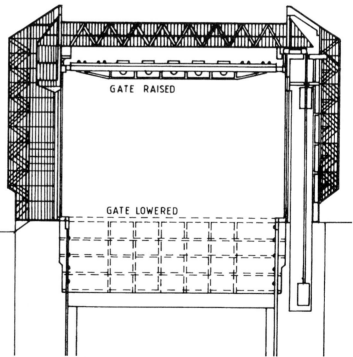

GATE RAISED

GATE LOWERED

Figure 2.37. River Hull tidal surge barrage

Figure 2.38. Vertically hinged sector lock gates for storm surge protection of Tokyo City

Actuation is by an electric-motor-driven winch and haulage chains passing over sprockets.

2.1.8. Pointing gates

Pointing gates are vertically-hinged, double-leaf flap gates. They operate automatically without external power on movement of the tide. They are used on small rivers to prevent the ingress of tidal water and provide protection against storm surges. Older gates were constructed of hardwood. New gates of this type have been constructed in steel. The gates open at low tide due to river flow and close on the rising tide. Sealing is by face contact with the masonry structure. The gates can slam on closure. Damage to masonry works sometimes occurs as a result of violent closure. No attempt appears to have been made to dampen this or to provide buffers on the gates in existence in Somerset in England.

In existing installations, pointing gates are backed by vertical-lift gates which provide a standby in case of breakdown of a gate or malfunction due to debris being caught between the gate leaves preventing complete closure.

2.1.9. Drum and sector gates

Drum and sector gates are acute circular sectors in cross-section. Gates hinged on the upstream side are referred to as drum gates (Figure 2.39 (a)) and those hinged on the downstream side as sector gates (Figure 2.39 (b)). The gates are designed so that they can be fully retracted to make the upper surface coincide with the crest line. Control of the gates is automatic and is by admission of the upstream water level into the float chamber.

Drum gates float on the lower face of the drum whereas sector gates are usually enclosed only on the upstream and downstream surfaces. These gates are not suitable for low dams because of the deep excavation required and the possibility of flooding of the float chamber due to downstream water level. Some very large drum gates have been built up to 40 m long and 9 m high.

Drum and sector gates have been superseded by radial gates at spillways because they are more complex to manufacture and therefore more costly. The cost of civil engineering works associated with drum and sector gates is significantly higher than with radial gates.

2.1.10. Bear-trap gates

A bear-trap gate consists of two leaves, one hinged upstream the other downstream. Both leaves are sealed at their side and pivots and are free to slide or roll relative to one another with a sliding seal at their juncture.

When the gate is lowered, the leaves come to rest in the horizontal position with the upstream leaf on top of the downstream one. When the upstream water level is admitted to the chamber 'a' the gate can be raised (Figure 2.40).

The water pressure under the gate is controlled either by an adjustable

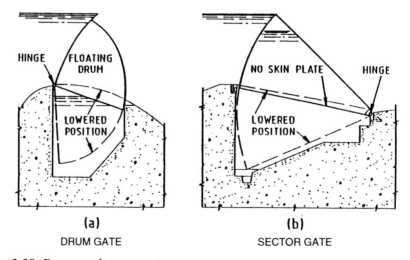

Figure 2.39. Drum and sector gates

weir or by setting inlet and outlet sluice valves in a control chamber in the sluiceway abutment.

Bear-trap weirs have been used in the US for log-sluicing operations, when the skin plates are usually protected by hardwood skid timbers.

The accumulation of silt under a bear-trap weir set on the river bed has been a source of trouble and various methods have been developed for the removal of silt by sluicing.

The seals of a bear-trap gate are critical as well as the control system. There is a recorded instance of a breakdown of a bear-trap weir due to vibration caused by bad design of the hinge seal. This problem will apply equally to bottom-hinged flap gates.

The calculations for each equilibrium condition of the gate have to be separately carried out, taking into account the external water load, the differential head required to raise the gate and the water level in the chamber to maintain the gate in position. The gate can be arranged to operate automatically to maintain upstream water level, although close control with variable tail-water levels is difficult to achieve. Raising the gate by admission of water to chamber 'a' requires effective side and sill seals. Bear-trap gates are now seldom constructed and when used are often raised by mechanical lifting gear travelling across the weir.

2.2. Gates in submerged outlets

2.2.1. Intake gates

Vertical-lift intake gates can be of the upstream or downstream sealing type as shown in Figure 2.17(a). Upstream sealing gates are located in a shaft, a short distance from the intake. Gates which control flow, or have to be shut against flow in an emergency, have to be operated by an oil hydraulic servo-motor. The reason why a gate on an elastic suspension such as a wire rope can be subject to vibration due to high velocity flow under the gate is

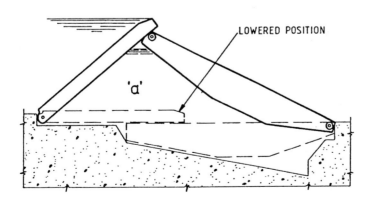

Figure 2.40. Bear-trap gate

discussed in Chapter 10. If the servo-motor is located above reservoir level, it has to be connected to the gate by a series of stem sections interconnected by knuckle joints, as shown in Figure 2.41.

The gate is raised to its maintenance position by hoisting it through the full stroke of the servo-motor, dogging the next lowest knuckle joint, followed by removing the uppermost stem and lowering the piston so that the piston rod can be reconnected to the next lowest stem. This operation is repeated until the gate is raised to its service position.

Downstream-sealing gates, and in some cases upstream-sealing gates, require air supply pipes which are extended to above reservoir level. Upstream-sealing gates are usually opened under conditions of balanced head, and this invariably applies to bulk-head gates. To effect balanced head conditions, either tunnel filling valves are incorporated in the gate or a valve-controlled bypass system is provided. In rope-suspended gates tunnel filling valves, integral with the bulkhead or intake gates as shown in Figure 2.42, are opened by initial tensioning of the hoist ropes. A short movement of the hoist lifts the valves. Tension in the ropes is sustained until the pressure on both

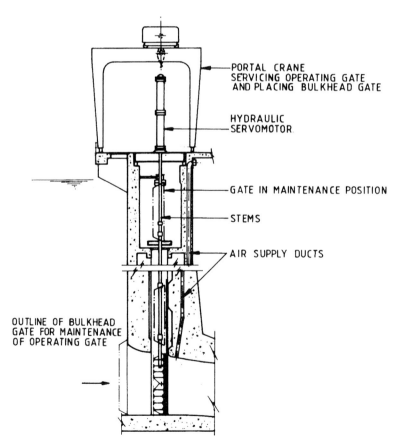

Figure 2.41. Intake gate servo motor operated

sides of the gate is equalised. Further hoisting of the ropes then lifts the gate. The type of valve suitable for this application is also used in stoplogs where it performs the same function and is illustrated under stoplogs in Figure 2.51.

Valves controlling bypass systems of high-head gates should be checked for cavitation conditions. In a piped bypass controlled by an operating and a guard valve, it may require a tapered outlet to cause a back-pressure, so that cavitation occurs external to the discharge section. Protection of the sluice wall from an impinging jet may also be required.

Intake gates which are not required to shut against flow, and bulkhead gates, can be rope suspended and lowered or hoisted by a conventional winch (Figure 2.42). Intake and bulkhead gates can be of the fixed roller type or for very heavy duty a caterpillar gate, known in the US as a coaster gate (Figure 2.43) is used.

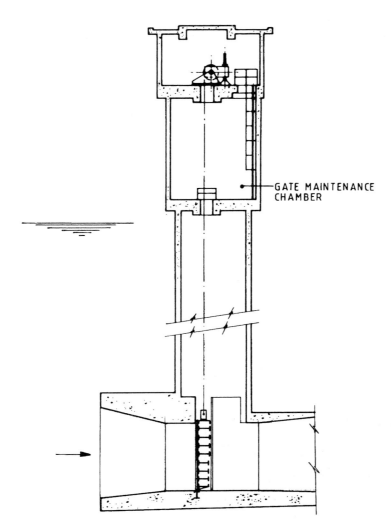

GATE MAINTENANCE CHAMBER

Figure 2.42. Intake gate, rope operated

Slide gates are also used as intake gates. If the gate is operated only under balanced conditions, the slide material has to be able to withstand the hydrostatic forces with no head on the downstream side of the gate. The frictional properties of the slide material in this application are of secondary importance. Impregnated woven asbestos is sometimes used. If the slide gate controls the intake flow, the slide bearing material is similar to that of control gates.

Intakes can be controlled by radial gates, as shown in Figure 2.44. The advantage of using radial gates is the absence of gate slots, which can cause hydraulic problems in high velocity flow, guide rollers which have to operate totally immersed, or slides. The gate is rigid with no slack in its movement and operating forces are less than those required for vertical-lift gates.

The disadvantages are a requirement for a chamber to retract the gate, that the gate cannot be withdrawn to the surface for maintenance and, in some cases, that the operating cylinder or cylinders are immersed while the gate is in the open position.

Cylinder gates (Figure 2.45) are used where the controlling gate must operate in a shaft or intake tower. They are used as shut-off gates and for regulating the intake. The gates are guided by rollers operating on tracks fixed to the tower walls, and therefore have little mechanical friction to overcome hydrodynamic excitation. Long operating stems or suspension chains of cylinder gates can result in low resonance frequencies.

The problem of gate vibration is dealt with in Chapter 10. Researchers have reported problems with cylinder gates which are specific applications of general principles. Vibration experienced at some cylinder gates appears to have been due to lack of a sharp cut-off point at lower lip. Vibration which has occurred at low gate openings is consistent with the variation of hydraulic downpull forces due to unstable flow.

 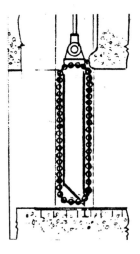

Figure 2.43. Coaster gate

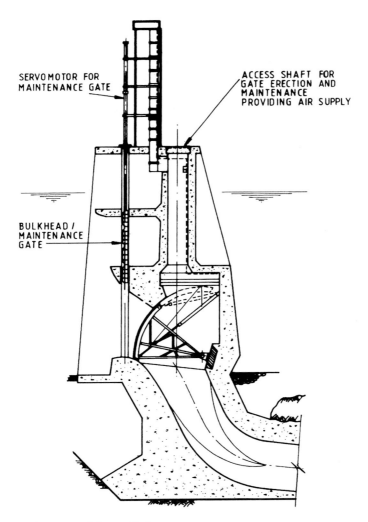

SERVOMOTOR FOR
MAINTENANCE GATE

ACCESS SHAFT FOR
GATE ERECTION AND
MAINTENANCE
PROVIDING AIR SUPPLY

BULKHEAD /
MAINTENANCE
GATE

Figure 2.44. Intake gate of the radial type

Ball[21] conducted a model study of a high-head cylinder gate which demonstrated that cavitation could occur due to the preliminary gate seat design and also vibration of the gate. Bixio *et al.*[22] found asymmetric pressure distribution on the shell of a cylinder gate due to unsteady flow through the eight openings of the intake. Negative pressures were recorded under emergency closure conditions, especially at the lower edges of the gate.

2.2.2. Control gates and guard gates

Control gates and guard gates can be vertical-lift, roller or slide gates which are servo-motor operated and retract into a bonnet (Figure 2.46 and 2.47). Frequently two identical gates are used.

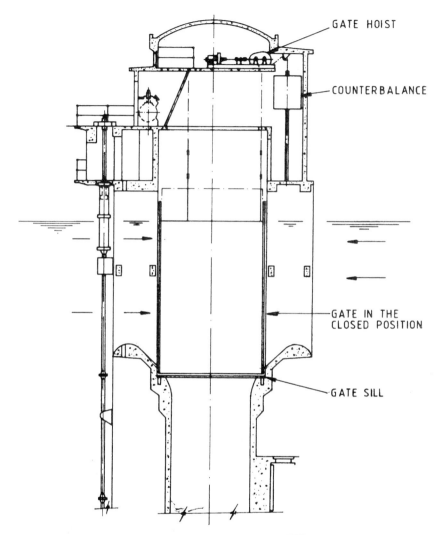

GATE HOIST

COUNTERBALANCE

GATE IN THE
CLOSED POSITION

GATE SILL

Figure 2.45. Intake gate of the cylinder type

The discharge from a slide gate is smooth and the only limitation is discharge at very small openings, when the flow does not spring clear of the lip of the gate and is liable to produce cavitation damage at the bottom of the gate.

The gates are designed with the skin plate downstream and with open stiffener girders on the upstream side. This causes flow circulation between the girders which is not detrimental. An alternative design is box construction with the space between the girder flanges filled in. Some gates, such as the bottom-outlet gate at the Victoria Dam in Sri Lanka, have been manufactured from solid forged steel plate. The bonnet is designed to withstand the full hydrostatic head without any structural contribution of the embedding concrete.

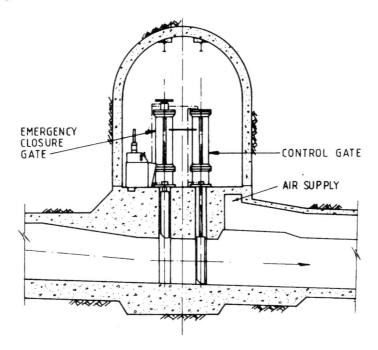

EMERGENCY
CLOSURE
GATE

CONTROL GATE

AIR SUPPLY

Figure 2.46. Control and emergency closure gates of the slide type

Figure 2.47. Slide gate

Unless the deflection of the gate is very low, it may be desirable to cross radius the bearing to allow for gate deflection, in which case the contact pressure (Hertz) calculations have to be carried out accordingly.

The slides can be conventional bearing materials such as leaded bronze, aluminium bronze or manganese bronze. Used on their own they require high-pressure grease lubrication. This has to be applied to the slide contact face within the gate slot by pipes leading from grease nipples at the top of the bonnet to a number of selected points on the slide face. Grease distribution grooves must be incorporated in the slide face to distribute the grease. When applying grease, effective distribution will occur only at grease outlets masked by the gate. There is a danger that grease or lubricant on exposed slide faces can be washed away by the recirculatory flow within the gate slots when the gate is in a partially or wholly open position. Possible environmental contamination may have to be considered. The use of bearings with lubricant inserts eliminates the sliding-seat greasing pipes and ensures an even lubrication coverage. In this type of slide, the lubricant is compressed into trepanned recesses in the bearing. The lubricant is of a permanent, solid, thick-film nature and is a compounded mixture of metals, metallic oxides, minerals and other lubricating materials combined with a lubricating binder. Graphite-containing lubricants should not be used in conjunction with stainless steel as they cause electrolytic action, which is accelerated under water.

When a gate is opened and discharges into an empty tunnel, an air demand is created due to air entrainment in the air water transition region. The calculation of air demand is dealt with in Chapter 9.

Difficulties can be experienced at gate slots at high velocity flow. In Chapter 9 these are discussed in more detail. Hydraulic problems in gate slots have led to the development of jet-flow gates. The gates incorporate contraction slopes on the conduit upstream from the gate slots to cause the flow to jump the slots in order to avoid intermittent flow attachment (Figure 2.48).

The jet-flow gate shown in Figure 2.48 is of the United States Bureau of Reclamation (USBR) circular-orifice type. Other types of jet-flow gate have been developed, having rectangular outlets. A rectangular gate seals flush at the sill and the contraction section is omitted at the sill. The rectangular jet-flow gate is appreciably cheaper to manufacture. The cost advantage is to some extent offset by a requirement for a transition section of conduit from circular to rectangular and on the downstream side from rectangular to circular.

Radial gates can be used as control gates in conduits. This can be arranged as shown in Figure 2.49, or the gate is located in a shaft as shown in Figure 10.16. As previously discussed, the main advantage of using a radial gate is the absence of gate slots. The disadvantages are a substantially larger gate chamber or shaft, and in many cases difficulties of access for initial assembly and subsequent maintenance.

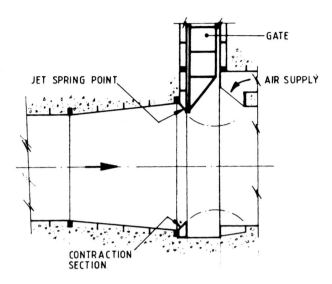

Figure 2.48. Control gate of the jet-flow type (circular orifice type)

The ring-follower gate is selected as a guard gate where the terminal-discharge is controlled by a valve. It avoids the upstream transition section from a circular section of conduit to a rectangular one, and a similar downstream transition section to a circular cross-section. Figure 2.50 shows a ring-follower gate. In the fully open position it provides an unobstructed fluidway. The gate can therefore reduce hydraulic losses to outlet works and result in economies in the transition sections.

The gate leaf retracts into the uppermost body, the bonnet section, when a circular opening aligns with the fluidway to present an unobstructed flow passage. To close the gate the circular opening is lowered into the bottom section of the bonnet and the bulkhead portion of the leaf blocks the fluidway. The lower bonnet has to be drained and designed for flushing of accumulated sediment.

The disadvantage of a ring-follower gate is its size, $3^{1}/_{2}$ times the diameter of the fluidway. This results in greater fabrication cost compared with a valve serving the same pipe diameter. However the size of a ring-follower gate is not limited, unlike that of a valve.

2.2.3. Emergency closure gates, maintenance gates and stoplogs

An emergency closure gate can close against flow. It will also perform the function of a maintenance gate. It requires load rollers. A maintenance gate will not normally close against flow. It can incorporate guide rollers or rely entirely on slides. It has to be placed under balanced conditions. Both emergency closure gates and maintenance gates can be in one section or in several sections assembled into one gate prior to lowering. Other terms are

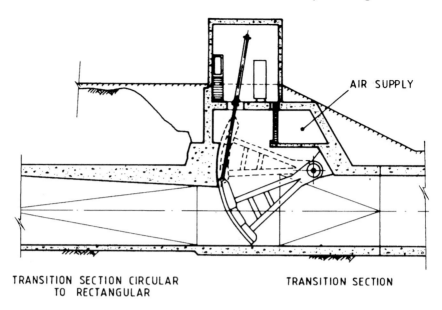

TRANSITION SECTION CIRCULAR   TRANSITION SECTION
        TO RECTANGULAR

Figure 2.49. Control gate of the radial type

in use such as bulkhead gate, guard gate or stop gate. In most cases these designate maintenance gates.

The term stoplog is a misnomer. It derives from the time when it was general practice to isolate a sluice installation by wooden beams. It would be more appropriate to designate them stop beams. Since the usage of the word stoplog is general it will be continued. Stoplogs cannot be placed in flowing water because they are liable to vibration during lowering and raising when combined over and underflow conditions occur. Stoplogs can be designed to be guided by rollers, or more frequently, by slides. Stoplogs incorporating guide rollers are easier to place.

Maintenance gates and stoplogs can be positioned by a rail mounted gantry crane or by a mobile crane. Emergency closure gates should only be placed by a rail mounted gantry crane. Hydraulic downpull forces (Chapter 9) could topple a mobile crane.

Figure 2.51 shows the essential features of a stoplog comprising lower seal 'a', upper seal contact plate 'b' and side seals 'c'. If the side seals are designed to seal both upstream and downstream it enables the gates to be tested hydrostatically on commissioning before the reservoir is fully impounded. If this is considered desirable the stoplogs must also be designed for reversal of the hydrostatic thrust. Bypass valves 'd' are linked with the grappling beam anchor points so that the initial lift movement of the grappling beam opens the valves to equalise the water level upstream and downstream of the stoplogs. Landing sensing device 'e' on the stoplog is positioned on the sillbeam or on top of another section. The rod is displaced

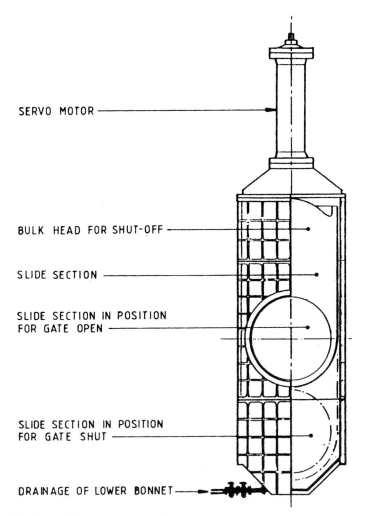

SERVO MOTOR

BULK HEAD FOR SHUT-OFF

SLIDE SECTION

SLIDE SECTION IN POSITION
FOR GATE OPEN

SLIDE SECTION IN POSITION
FOR GATE SHUT

DRAINAGE OF LOWER BONNET

Figure 2.50. Ring-follower gate (half section)

upwards and permits disengagement of the grappling beam. If the stoplog gets jammed, or meets an obstruction, release of the grappling beam cannot be actuated or accidentally effected.

Stoplog guide channels

The same structural design criteria apply to the design of stoplog guide channels as to those of vertical-lift gates. They are usually simpler with embedded parts only for sliding or rolling face. If it is important to enhance the hydraulics of the flow through the sluiceway, stoplog masking plates are used. They are placed and withdrawn by the crane handling the stoplogs.

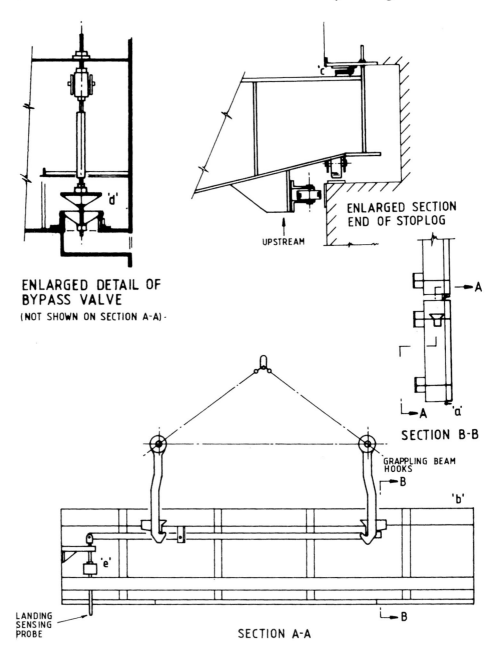

ENLARGED DETAIL OF
BYPASS VALVE
(NOT SHOWN ON SECTION A-A)

ENLARGED SECTION
END OF STOPLOG

UPSTREAM

SECTION B-B

GRAPPLING BEAM
HOOKS

LANDING
SENSING
PROBE

SECTION A-A

Figure 2.51. Stoplog section

Grappling beams

Figure 2.52 shows a typical grappling beam. It is designed so that it will automatically engage with a stoplog and when the beam is in position and the landing sensing device is activated it will automatically disengage. The lever 'a' has to be moved to determine whether the grappling beam will engage or disengage. Guiding in transverse and the longitudinal direction is effected by rollers.

Cranes

Emergency closure gates must be handled by a rail-mounted gantry crane, whereas maintenance gate and stoplog handling is either carried out by a rail-mounted gantry crane or a mobile crane. Gantry cranes can be used to transport maintenance gates and stoplogs from their storage area to the sluiceway whereas a mobile crane cannot, as a rule, transport heavy gates or stoplogs. If this is the case two small rail bogies can be provided.

Different functions are often carried out by the same crane. In Figure 2.41 it can service the operating gate and place the maintenance gate. In other installations it is used for servicing the operating gate and as a means of placing stoplogs or an emergency closure gate. When the crane has a cantilever runway with an auxiliary hoist, it can also raise a removable screen of the type shown in Figure 4.2. In a limited number of intakes, mainly freestanding in a reservoir and accessed by a bridge, the gantry crane servicing the operating gate can also place the emergency closure gate and mounts integrally raking machinery, carrying out the same functions as the equipment illustrated in Figure 4.3. Creep speeds on all motions of a gantry crane should be provided to enable accurate positioning. Creep speeds should be about $1/10$ to $1/15$ of normal motion speeds.

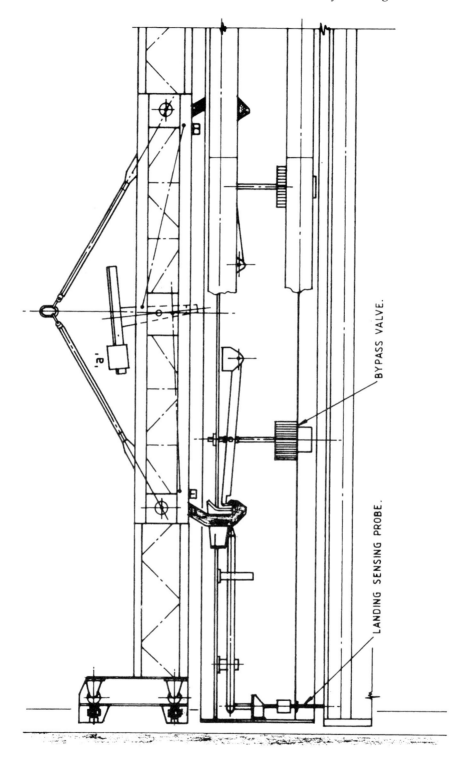

Figure 2.52. Grappling beam and stoplog

Appendix 2.1. Main applications, advantages and disadvantages of various types of hydraulic gate

Type	Main Applications	Advantages	Disadvantages

Gates in open channels

Type	Main Applications	Advantages	Disadvantages
1. Radial gates, motorized	Sluice installations. River control. Spillways. Barrages.	No unbalanced forces. Absence of gate slots. Low hoisting force. Mechanically simple. Bearings out of the water. Can be fitted with overflow section. Some inspection with gate in service possible.	Extended flume walls. High concentrated loads. Increased fabrication complexity.
2. Radial automatic gates	Sluice installations. River control.	No outside source of power required. Absence of machinery. Low maintenance.	Wide piers to accommodate displacers. Counterbalance visually intrusive. Can malfunction due to incorrect design. Can malfunction due to blockage of inlet of control system.
3. Vertical-lift gates	Sluice installations. River control. Old installations: Barrages Spillways.	Can be fitted with overflow sections. Short piers. Wide-span gates can be engineered to provide good navigation openings. Up-and-over gates can reduce height of supporting structure.	Gate slots required. Load rollers under water. Can jam due to debris. High hoisting load unless counterbalanced. Overhead support structure visually intrusive.
4. Flap gates, bottom-hinged	Tidal barrages. Sluice installations. River Control.	Complete separation of saline and fresh water. Overflow to clear debris. No visually intrusive overhead structure Can in some cases be engineered to open under gravity in an emergency.	Requires extensive side staunchings for side sealing or very accurately constructed pier walls. Hinge bearings not easily accessible and permanently immersed.
5. Flap gates, top-hinged	Tidal outlets.	Requires no outside source of power. Automatic in operation. Simple construction. Little maintenance required.	Cannot control water level. Will not entirely exclude tidal water if downstream water level rises above sill. Gate slam can occur.

Type	Main Applications	Advantages	Disadvantages

Gates in open channel—continued

Type	Main Applications	Advantages	Disadvantages
6. Rising-sector	Storm-surge barriers.	Unobstructed navigation passage Not intrusive visually. Can be raised for maintenance and inspection without stoplogs.	Requires piers. Permits some flow upstream. Complex fabrication of gate. Complex machinery.
7. Buoyant gates, bottom-hinged	Storm-surge barriers. Tidal barriers.	Unobstructed navigation passage without piers. No structure above navigation passage bed level. Excellent from visual considerations.	Capable of withstanding only limited differential head. Gates move independently under wave action resulting in leakage between gates. Requires detachable hinges. Gates have to be interchanged for maintenance.
8. Mitre gates	Storm-surge barriers.	Unobstructed navigation passage. Excellent from visual considerations.	Width of navigation more limited than some other gates. Closure and opening has to be effected when water levels are nearly equal. Heavy mitre thrust. Cannot open or close against flow.
9. Vertically-hinged sector	Storm-surge barriers.	Unobstructed navigation passage. Can be opened or closed against flow. Can be opened at differential head. Can be constructed with wider opening than mitre gates. Excellent from visual considerations.	Requires wide recess in the banks to accommodate gates on opening.
10. Drum and sector gates	Spillways.	Requires no outside source of power	Complex gates. Requirement for extensive civil engineering works. Requires zero downstream level. Control system critical. Can silt up. Not favoured.

Type	Main Applications	Advantages	Disadvantages
Gates in submerged outlets			
1 Vertical-lift intake gate, servo-motor operated	Control and emergency closure.	Reliable control gate. Good load distribution in the slide version. Damped.	Gate slot required. Load rollers or slides operating under water. Requires stem connections between servo-motor and gate. Possible cavitation problems. Slow operation to raise to the maintenance position. Requires air admission.
2. Vertical-lift intake gate, rope operated	Bulkhead gate.	Can be load roller or slide gate. Does not require air admission.	Cannot be used as a control gate. Cannot be used as an emergency closure gate. Requires balanced head for operation. Guide slots required. Possible cavitation problems. Requires bypass system.
3. Caterpillar or coaster gate	Control and emergency closure gate.	Control gate for very high heads.	Wide gate slots required. Caterpillar train operates under water. Requires stem connections between servo-motor and gate. Cavitation problems. Slow operation to raise to the maintenance position. Very costly. Requires air admission.
4. Radial-intake gate	Control and emergency closure gate. Intake gate.	Absence of gate slots. Requires no load rollers or slides.	Requires chamber to retract. High concentrated load. Lintel seal critical. Requires dewatering of tunnel to carry out maintenance. Requires air admission.

Type	Main Applications	Advantages	Disadvantages
	Gates in submerged outlets—continued		
5. Cylinder gates	Intake gate.	Capable of controlling intake flow and large openings.	Low natural frequency of vibration due to rope suspension and low friction. Possible vibration problems. Large gates require counter-balance to reduce hoisting forces.
6. Slide gates	Control gates in conduit. Back-up gate for a control gate.	Reliable control gate or emergency closure gate. Inherently damped due to sliding friction.	Gate slot required. Possible cavitation problems. Requires bonnet for withdrawal. Requires air admission.
7. Jet-flow gates	Control gates in conduit for high-head application.	Reliable control gate at high heads. Inherently damped due to sliding friction. Can be circular or rectangular.	Gate slot required. Requires bonnet for withdrawal. Requires air admission.
8. Radial gates	Control gates in conduit.	Absence of gate slots, rollers or slides. Lower hoist force required compared with 6 and 7.	Requires chamber to retract. High concentrated load. Lintel seal critical. Requires dewatering of tunnel to carry out maintenance. Requires air admission.
9. Ring-follower	Back-up gate for terminal discharge gate or valve.	Can control flow. Provides unobstructed flow. Does not require transition section from circular to rectangular duct.	Gate slots required. Large overall height—approximately three times of fluid way. Requires regular flushing. Drain connection must be provided.

References

1. Rouse, H (1964): Engineering Hydraulics, *Proc. 4th Hydr. Conference*, Iowa Institute of Hydraulic Research, Jun 1949, John Wiley & Sons Inc.
2. Murphy, T E (1963): Model and Prototype Observation of Gate Oscillations, *10th I.A.H.R. Congress, London*, paper 3.1.
3. Lewin, J (1983): Vibration of Hydraulic Gates, *J. I.W.E.S.*, 37, 165.
4. Thorne, R B (1957): The Design, Fabrication and Erection of Radial Automatic Sluice Gates, *Proc. I.C.E.*, 6th Feb, 126–133.

5. Lewin, J (1984): Radial Automatic Gates, *Proc. 1st Int. Conference Channels and Channel Control Structures*, Southampton, paper 1–195, editor Smith, K V H, Springer Verlag.

6. Lewin, J (1991): *Gates for Estuarial Outlets*, I.W.E.M., River Eng. Section, Leamington Spa, May.

7. Randerson, R J (1979): A Velocity Control Structure in the River Orwell, Ipswich, *J.I.W.E.S.* 38, 135.

8. Petrikat, K (1958): Vibration Tests on Weirs, Bottom Outlet Gates, Lock Gates, *Water Power*, Feb, Mar, Apr and May.

9. Naudascher, E (1965): Discussion on Nappe Oscillation, *Proc. A.S.C.E., Journ. Hydr. Div.*, May.

10. Schwartz, H I (1964): Nappe Oscillation, *J. Hydr. Div., Proc. A.S.C.E.*, HY6, Nov, paper 4138.

11. Partenscky, H W; Swain, A (1971): Theoretical Study of Flap Gate Oscillation, *14th I.A.H.R. Congress, Paris*, Paper B.26.

12. Krummet, R (1965): Swingungsverhalten von Verschlussorganen im Stahlwasserbau, *Forschung im Ingenieurwesen*, Bd. 31, No. 5.

13. Falvey, H T (1979): Bureau of Reclamation Experience with Flow Induced Vibrations, 19th *I.A.H.R. Congress, Karlsruhe*, paper C2.

14. Ogihara, K; Ueda, S (1979): Flap Gate Oscillation, *19th I.A.H.R. Congress, Karlsruhe*, paper C11.

15. Pulpitel, L (1979): Some Experiences with Curing Flap Gate Vibration, *19th I.A.H.R. Congress, Karlsruhe*, paper C12.

16. Nielson, F M; Pickett, E B (1979): Corps of Engineers Experience with Flow Induced Vibrations, *19th I.A.H.R. Congress, Karlsruhe*, paper C3.

17. Lewin, J; Scotti, A (1990): The Flood Prevention Scheme of Venice: Experimental Module, *J. Instn. Wat. & Envir. Mangt.*, 4,1, Feb.

18. Hamilton, A J; Prosser, M J (1988): *Venice Lagoon Flood Protection, Hydraulic Model of Scouring System*, B.H.R.A. report RR 2918.

19. Ayres, D (1983): The Thames Barrier: the Background and Basic Engineering Requirements, in I. Mech. E *Seminar Proceedings The Thames Barrier*, 8th Jun.

20. Clark, P J; Tappin, R G (1977): Final Design of Thames Barrier Gate Structures, In. Conference on *Thames Barrier Design*, London, Oct, I.C.E., London, 1978.

21. Ball, J W (1959): Cavitation and Vibration Studies for a Cylinder Gate Designed for High-heads, *8th Congress I.A.H.R., Montreal*, paper 9A.

22. Bixio, V; Cola, R; Garbin, C; Mariani, M (1985): On the Hydraulic Behaviour of a Cylinder Gate in a Vertical Intake with Radial Symmetry Openings, *2nd Int. Conference on the Hydraulics of Flood and Flood Control, Cambridge*, paper D3.

3
Valves

The flow in pipelines is controlled by valves. This chapter is mainly concerned with large valves carrying out the function of terminal-discharge and providing backup in circular conduits. The exceptions are pressure-reducing valves in pipelines, one example of which is shown in Figure 3.16, and the needle valve which can be used for regulating flow in pipelines or as a terminal-discharge valve. It is largely superseded as a discharge valve.

Appendix 3.1 lists the main applications and the advantages and disadvantages of the various types of valve.

In terminal-discharge valves, the hollow-cone valve predominates because of its good energy dissipation characteristics, its simplicity of construction, lower cost and favourable coefficient of discharge. It is also least prone to blockage.

Hollow-jet valves and needle valves with their internal moving parts, compared with the external control sleeve of the hollow-cone valve, are mechanically more complex and therefore more difficult to service. The fluid passages are more restricted and liable to blockage.

Trashracks and screens are essential at intakes which serve conduits containing valves.

3.1. Sluice valves

Sluice valves are the most frequently used control devices in pipelines. In the open position they provide an unobstructed fluid passage. In spite of their non-linear flow control characteristics they are often used for that purpose. Because the valve blade is unsupported during raising and lowering and due to eddy shedding from the blade tip they are suitable only for closure and opening against flow at low velocities.

3.2. Butterfly valves

The butterfly valve is the most frequently used closure device in pressure conduits because of its relatively compact arrangement and simple construction. In the open position the blade lies in the plane of flow of the fluid. Valves are manufactured in sizes up to 4 m diameter and are able to withstand operating heads up to 200 m within that range of size.

There are two types of butterfly valve. The solid-disk valve, sometimes referred to as lenticular (Figure 3.1(a)) and the lattice-blade valve, also described as a through flow valve (Figure 3.1(b)). The latter offers the advantage of a stiffer disk assembly and lower loss coefficient[1.]

Valve blades are normally manufactured in cast iron or carbon steel. Other more corrosion resistant materials such as high nickel cast iron, stainless steel or aluminium bronze are used as the material for the blade where the water carries bed material or is aggressive to cast iron or steel. Valve bodies and blades have been coated for special applications with epoxy resin or ebonite.

Figure 3.1. Solid disk and throughflow butterfly valves

Figure 3.2 shows different arrangements of seals for butterfly valves. Seal (a), an all-metal seal, is suitable only for low operating heads. At higher pressures an elastomeric seal of Neoprene or Nitrile is used (b) and the seal is pressurised by the upstream water head. A diaphragm seal is shown in section (c). This seal is also pressurised by the upstream water.

The seal seat for (a) and (b) is of stainless steel and is welded into the housing. It is arranged flush and not as shown for clarity as projecting.

In the closed position the disks of the versions shown in sections (a) and (b) are inclined at approximately 80° at right angles to the conduit axis. This applies also to lattice-blade valves. The latter can be provided with a second seal on the upstream side which can be normally or hydraulically-operated, which allows replacement of the downstream seal while the conduit remains under pressure.

Butterfly valves are normally opened under balanced conditions and closed against flow. They are not in general suitable for flow control, only as on/off devices, because of flutter of the blades and eddy shedding from the blade tips. Prior to the opening of the valve, the pressure is balanced by means of a bypass pipe which incorporates shut-off valves. At high-heads manually, electrically or oil pressure actuated filling nozzles are used. A

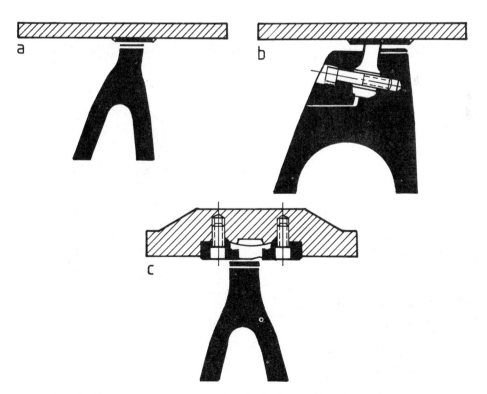

Figure 3.2. Different arrangements of seals for butterfly valves

guard valve is arranged upstream of the filling nozzle. Means of venting the conduit downstream of the valve during the filling operation must be provided and also during draining.

Closure of valves controlling conduits or penstocks is usually by gravity (Figure 3.3(a)). During normal operation the lever arm for the falling weight is locked in the valve open position. The locking mechanism is designed to release the arm which is then able to rotate to the shut position. The release mechanism may be triggered manually, electrically or indirectly by excessive velocity of flow in the conduit or by loss of pressure. The valve is opened by the oil hydraulic servo-motor which controls also the rate of closure. Shortly before the end of the closure movement the discharge of oil from the

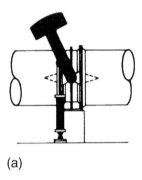

(a)

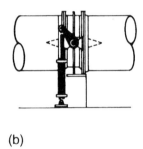

(b)

Figure 3.3. Operation of butterfly valve
controlling conduits or penstocks
(a) Gravity closure by falling weight
(b) Closure by double acting servo-motor

annulus side of the cylinder is throttled to ensure slow final closure of the valve. The closing time has to be determined so as to minimise the water hammer in the penstock. This can be effected also by cushioning the last part of the movement of the piston. Ellis and Mualla[2] analysed the closure characteristics of butterfly valves.

The servo-motor in Figure 3.3(b) is double acting. Opening of the valve is effected by oil or an emulsion supplied to the piston side of the cylinder from an hydraulic power pack while the fluid on the annulus side is ported to discharge. For closure of the valve, mains pressure is admitted from upstream of the butterfly valve to the annulus side of the cylinder while the oil is ported to the tank of the power pack. The water used as a closing medium is filtered before it reaches the control elements and the servo-motor.

Under conditions of emergency closure, cavitation will occur where the conditions of initial operation with positive back-pressure change into one of complete separation of flow. The time is usually short enough not to cause any damage.

The loss coefficient for fully open butterfly valves is shown in Figure 3.4. The loss coefficient of a valve K_v is defined as:

$$K_v = \frac{\Delta H}{v^2/2g}$$

where ΔH =   total head loss
        v =   velocity of flow in the valve
        g =   acceleration due to gravity

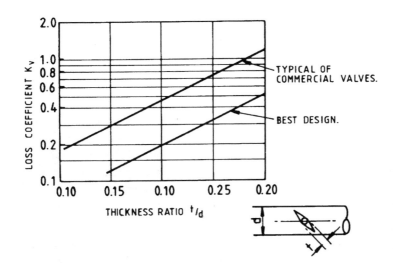

Figure 3.4. Loss coefficients for fully open butterfly valves (after Miller[3])

The loss coefficients for partially open valves are shown in Figure 3.5. Variations of more than 10% occur particularly when the valve is nearly closed and when the valve seating arrangement becomes very important (Miller[3]).

3.3. Cavitation in valves

Cavitation is caused by the local pressure on the downstream side of a valve by the accelerated flow of the water as it contracts to pass through the valve opening and by the generation of turbulence. Eddies are formed in the intense shear layer that surrounds the accelerated flow of water through the valve opening. When the pressure inside the eddies, which is considerably less than the fluid pressure in the penstock, approaches the vapour pressure, cavitation bubbles grow from nuclei suspended in the water. As the ambient pressure increases and the eddies degenerate by viscous forces, the bubbles become unstable and collapse. If this occurs next to a solid boundary, it creates noise, vibration and in its intense form erosion damage. The cavitation index can be used to express the level of cavitation as the ratio of forces suppressing or preventing cavitation to the forces causing cavitation.

$$\text{Cavitation index} = \sigma = \frac{hd - hv}{hu - hd}$$

$$\text{or } \frac{hu - hv}{hu - hd}$$

$$\text{or } \frac{hu - hv}{v^2 / 2g}$$

where hd = head downstream of the valve
       hv = vapour head at the inlet temperature
      hu = head upstream of the valve
        v = average pipe velocity

Note: Consistent pressures either absolute or gauge must be used. If gauge pressures are used the vapour head has a negative value.

The higher the cavitation index the least likely is cavitation damage. Chapter 9 discusses the different intensities of cavitation—incipient, critical and choking cavitation. Increasing from the initial stage, incipient, which consists of light intermittent bursts of noise, to choking cavitation when intense noise and severe damage occur. Rahmeyer[4] carried out measurements of cavitation intensity in valves and the American Instrument Society[5] lays down a test procedure. Miller[3] gives a graph of incipient, critical and choking cavitation for v and K_v which is reproduced in Figure 3.6.

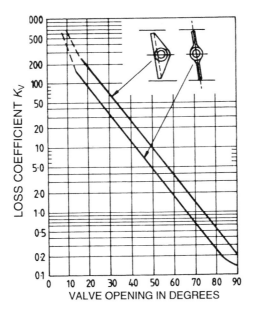

Figure 3.5. Loss coefficients for partially open butterfly valves (after Miller[3])

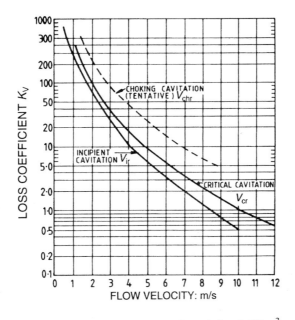

Figure 3.6. Cavitation velocities for butterfly valves (after Miller[3])

Table 3.1. Vapour pressure of pure water

	Vapour Pressure	
°C	N/m^2	mm head of water
0	610	62
5	870	89
10	1230	125
15	1700	174
20	2330	238
25	3160	323
30	4230	433

The base conditions for the graph are a valve diameter of 0.31 m and an upstream head minus vapour head of 50 m. To correct these velocities to other valve sizes and head the following equation must be used:

$$V_{ir} \text{ or } V_{cr} = C_1 V_r \left(\frac{hu - hv}{50} \right)^{0.39}$$

where V_{ir} = incipient cavitation
        V_{cr} = critical cavitation
        C_1 = a correction factor
        V_r = velocity V_{ir} from Figure 3.6 for incipient cavitation or the reference velocity V_{cr} from Figure 3.6 for critical cavitation
C_1 is taken from Figure 3.7.

3.4. Hollow-cone valves and hoods

The valve is more commonly known by the surnames of the inventors, Howell and Bunger. It is widely used as a regulating valve for free discharge because of its simplicity. Figure 3.8 shows a cross-section of a hollow-cone valve. The figure shows a typical cross-section through the valve. The upper section is the closed valve and the lower one the fully open position. The valve body is cylindrical, flanged at the upstream end for attachment to the pipeline and connected to a downstream dispersing cone by streamlined radial ribs forming an annular outlet port. Flow control is effected by a reinforced stainless steel cylindrical gate which slides over gunmetal bearing strips, secured to the body, to close the annular port and to seal against a seat ring attached to the dispersing cone, which forms the discharging jet of water into a hollow divergent cone in which the energy of the jet is dissipated by air friction and entrainment.

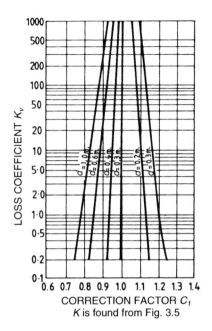

Figure 3.7. Correction factors for valve size

The discharge of the valve is given by:

$$Q = CdA\sqrt{(2gH)} \qquad\qquad 3.1$$

where Q = discharge
Cd = discharge coefficient (approximately 0.85)
A = area of the valve based on the inside diameter of the valve body
g = gravitational constant
H = net head at the valve entry

Westinghouse quote a Cd value of 0.85 for their valves. A study of a 2.5 m diameter valve carried out by Boving & Co. gave a value of 0.83.

Hollow-cone valves are manufactured in sizes up to 3.5 m diameter with operating heads up to 250 m. Two major types of hydrodynamic problems have been experienced with hollow-cone valves, vane failure and shifting of the point of flow attachment.

Vane failure

This has been attributed to a number of causes but the most likely one is hydroelastic instability causing vibration normal to the chord of the valve and twisting about the longitudinal axis. Destructive resonance occurs at a critical velocity at which the flow couples the two forms of vibration in such a way as to feed energy into the elastic system. Possible modes of vibration for

VALVE IN CLOSED POSITION

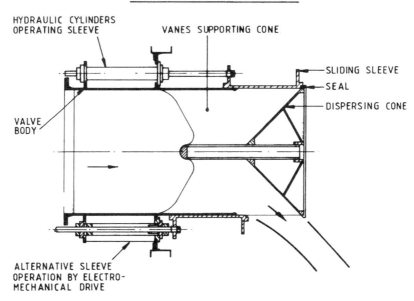

VALVE IN OPEN POSITION

Figure 3.8. Hollow cone valve

a hollow-cone valve are shown in Figure 3.9. Mercer[6] suggests a parametric value incorporating a coefficient depending on the ratio of shell to vane thickness and number of vanes. Valves with a value less than 0.115 have operated successfully and valves with a value greater than 0.130 have failed.

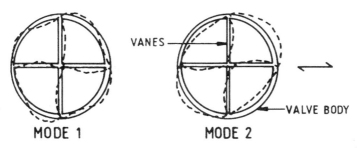

Figure 3.9. Possible vibration modes of hollow-cone valves (after Mercer[6])

Mercer's parametric value is:

$$\frac{Q/(CDT_v)}{\sqrt{(E/e)}} \qquad 3.2$$

where Q = discharge
      C = a dimensionless coefficient depending on K and N, the number of vanes
      D = valve diameter
      T_v = vane thickness
      E = Young's modulus
      e = density of fluid
      K = ratio of shell thickness to vane thickness = T_s/T_v

Mercer's investigation and tabulation of values was carried out in Imperial units and therefore consistent Imperial units must be used for equation 3.2 (Table 3.2).

The frequency of vibration f of mode 1 (Figure 3.9) can be expressed by the equation:

$$f = C\sqrt{[(E/e)/(T_v/D^2)]} \qquad 3.3$$

Inserting the value of C in this equation shows that six-vane valves have 10% higher frequencies than comparable four-vane valves and that the thickness of the shell relative to the vane does not have too large a bearing on the frequency.

Nielson and Pickett[7] reported a major vane failure of a 2740 mm diameter

Table 3.2. Values of C in equation 3.2

N	4	5	6	6	6	6	6
T_s/T_v	1.00	1.00	0.50	0.90	1.00	1.20	2.00
C	2.22	2.35	1.98	2.40	2.48	2.53	2.75

hollow-cone valve. The failure was of the fatigue type. Mercer's parametric value of the valve as originally constructed was 0.176.

Falvey[8] cited severe vibration of two 2135 mm hollow-cone valves. The observed 85 Hz frequency correlated well with estimates of its natural vibration frequency based on the paper by Mercer. The valve opening in the prototype had to be restricted to a maximum of 80%.

Shifting of the point of flow attachment

As the hollow-cone valve is opened the flow control may shift from the sleeve to the valve body (Figure 3.10) and intermittent attachment and re-attachment may occur resulting in severe vibration[7]. Under these conditions the opening of the valve, that is the sleeve travel, has to be limited.

Deterioration of the seal between the valve body and the sliding sleeve can result in leakage. In general this does not lead to vibration of the valve but if it continues for a long time it can result in erosion.

The expanding cone-shaped discharge pattern of the hollow-cone is very effective in aerating the water and dispensing the energy. Because of these features, stilling basins are not normally used for the discharge of hollow-cone valves. The usual angle of the cone is 45°. Experimental investigations have been carried out with valves having a cone of different angles[9].

Where hollow-cone valves are located in a tunnel or where the spray from a widely dispensed jet is not acceptable, a hood is used to confine and redirect the discharge as shown in Figure 3.11. In tunnels or conduits the hood prevents erosion by the impinging jets and ensures that air is admitted from upstream. Guidelines as to the optimum geometry of hoods are given by Brighouse and Chang[10,11]. The hood has to be arranged to minimise

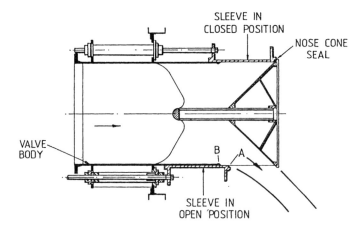

Figure 3.10. Vibration of hollow-cone valve due to the shifting of the point of flow attachment—oscillating between A and B (after Nielson and Pickett[7])

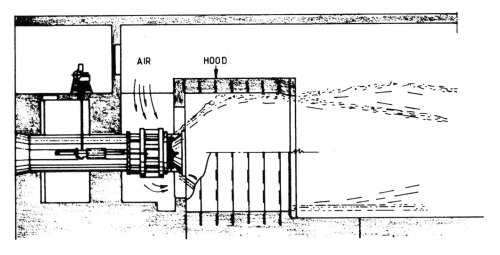

Figure 3.11. Installation of a hollow cone valve with hood

splashback through the upstream opening and ribs have to be introduced on the inside of the discharge section of the hood so that air is admitted to the inside of the discharge jet. Hollow-cone valves have been used submerged within a stilling basin. In this application they are not energy dissipating devices and this function is carried out by sheared flow and air which is introduced into the stilling basin. A model study is required for the successful installation of submerged hollow-cone valves. Figure 3.12 is an illustration of such an application. Flotsam can get wedged across the vanes of a hollow-cone valve but is usually only a problem in smaller valves.

Hollow-cone valves can be installed to discharge into a stilling basin or submerged as shown in Figure 3.12. In the former case the main purpose of the valve to act as an energy dissipator is limited because the jet is shortened and therefore its ability to entrain air.

3.5. Hollow-jet valves

Figure 3.13 illustrates the construction of a hollow-jet valve. Movement of the cone controls the area of the discharge orifice. It is used as a terminal-discharge and control valve. The jet is compact and therefore entrains less air than the hollow-cone valve. It can be installed directly after a bend in the pipework.

The hollow-jet valve is frequently installed so that it discharges at an angle of 30° into a stilling basin. The flow in the conduit by the moveable cone and the body results in a hollow-jet, which initially returns to its shape and flares out shortly before the point of impingement. A jet angle relative to the horizontal loosens up the jet structure and reduces the intensity of impingement.

Inspection and servicing of the mechanical or the oil hydraulic actuator

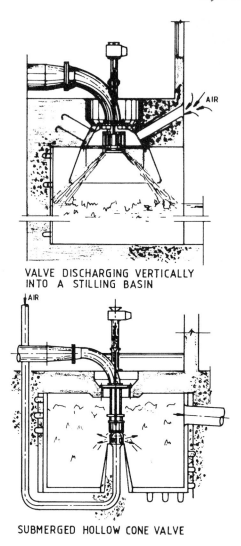

VALVE DISCHARGING VERTICALLY
INTO A STILLING BASIN

SUBMERGED HOLLOW CONE VALVE

*Figure 3.12. Installation of a submerged hollow cone valve and valve
discharging vertically into a stilling basin*

requires removal of the complete valve. Operation of the oil hydraulic
operated valve (Figure 3.13(b)) is by an external hydraulic power pack with
pipework which has to be routed through the jet.

The coefficient of discharge of a hollow-jet valve is about 0.7 at full valve
opening, reducing to 0.4 at the half-open position and 0.23 at quarter
opening.

Hollow-jet valves are manufactured in similar range of sizes and head as
hollow-cone valves.

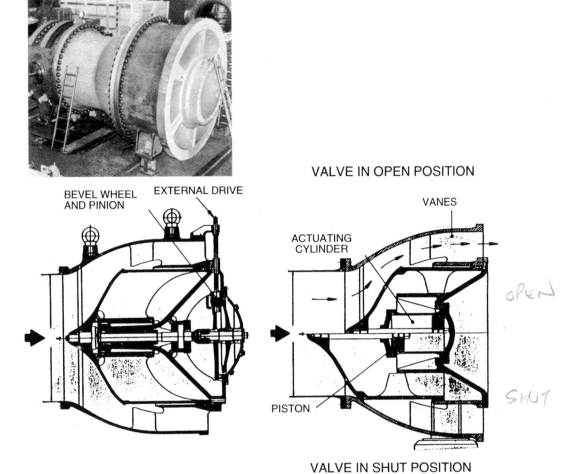

a) Section through a hollow jet valve with mechanical actuator

b) Section through a hollow jet valve with oil hydraulic actuator

Figure 3.13. Hollow jet valve

3.6. Needle valves

Needle valves are used for regulating flow, either as terminal-discharge valves or for controlling high-head flow in pipes. Their use in outlet works has been supplanted in many applications by more economical and hydraulically efficient valves, such as the hollow-cone valve. Figure 3.15 shows an installation and cross-section through an interior-differential needle valve. The valve is closed by admitting water pressure to chamber B and connecting chamber A to drain through the spool valve located at the bottom of the needle valve. To open the valve, water pressure is admitted to chamber A and chamber B is opened to drain.

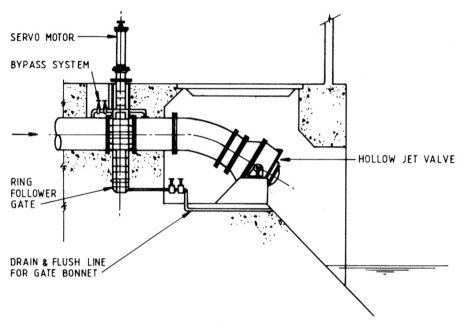

Figure 3.14. Installation of hollow jet valves

To prevent cavitation the discharge opening is arranged so that the downstream cone angle of the needle is slightly less than the downstream cone angle of the body. A sharp flow separation point at the body seat is another requirement if cavitation is to be avoided. The coefficient of discharge using equation 3.1 is about 0.6 at full valve opening. This reduces to 0.4 at half opening and 0.26 at quarter opening. Because of the low coefficient of discharge at partial openings, the valve can dissipate energy when controlling high-head flow in pipes. Needle valves are manufactured in sizes up to 2 m and for working heads up to 200 m.

3.7. Pressure-reducing valves

Figure 3.16 illustrates a pressure-reducing valve. The energy dissipation is effected by discharging some or all of the flow through the orifices of the perforated cylinder. The cylinder is arranged on a plunger which advances or retracts the throttling cylinder. The drive can be manual or be powered by an electric actuator driving the plunger via a bevel gear. The perforations break up the flow into numerous concentric individual jets directed against one another. The valve is suitable only as a regulating valve in closed pipe systems. It is also used as a bypass valve. It is manufactured in sizes up to 1.5 m diameter.

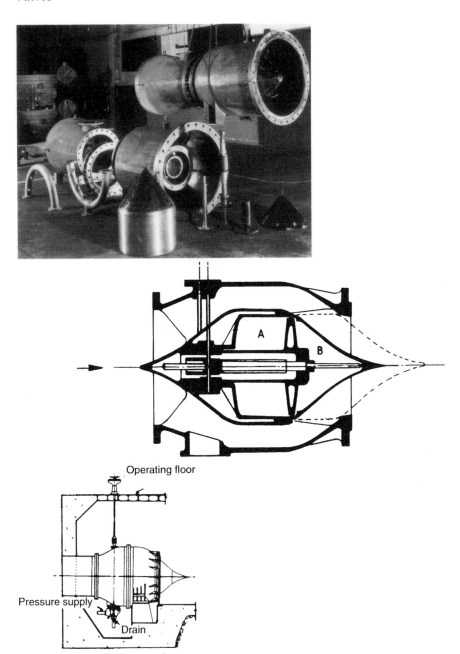

Figure 3.15. Interior-differential needle valve

3.8. Sphere valves

Figure 3.17 shows a sphere valve, sometimes denoted as a rotary valve. The section shows the lower half of the valve fully open and the upper half fully closed.

Right: Perforated throttling cylinder in closed position
Left: Perforated throttling cylinder in half-open position

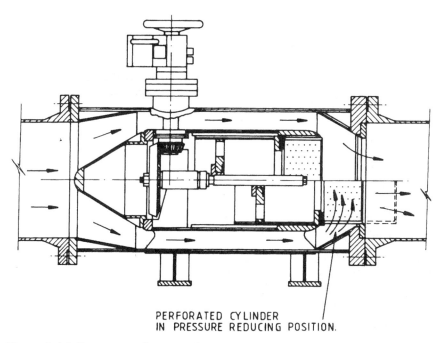

PERFORATED CYLINDER
IN PRESSURE REDUCING POSITION.

Figure 3.16. Pressure-reducing valve

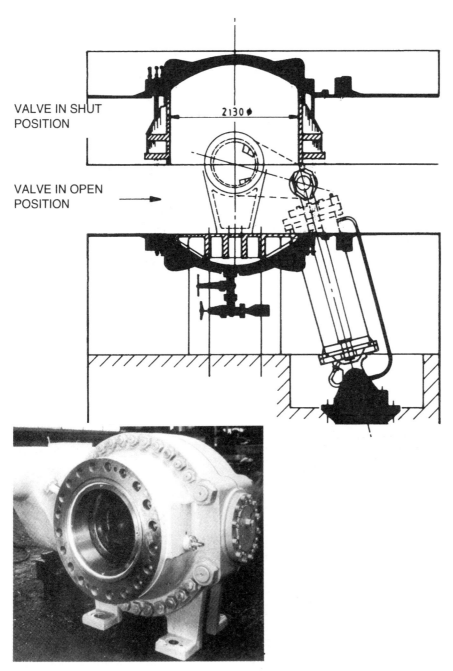

VALVE IN SHUT
POSITION

2130 ⌀

VALVE IN OPEN
POSITION →

Figure 3.17. Sphere valve

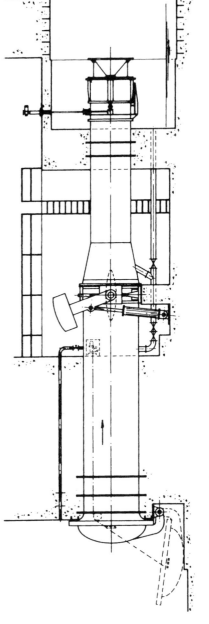

Figure 3.18. Arrangement of bottom outlet valves: butterfly valve and hollow cone valve

Sphere valves have a clear bore and when fully open have a very low loss coefficient. Resilient rubber seals are used for valves working at pressures up to 400 m head. For higher heads metal seals are used. Valves are normally supplied with an operating seal at the downstream end and an additional manually or hydraulically-operated maintenance seal on the upstream end. Closure is droptight. The application of sphere valves is for shut-off control on the pressure side of high-head turbines and pumps. The usual arrangement is gravity closure of the valve and oil hydraulic piston operation for opening. Alternative operations of the servo-motor have been used where the opening is effected by hydraulic oil acting on the piston side of the cylinder and uncontrolled operating water on the annulus side. On opening the hydraulic oil on the piston side overcomes the force on the water side displacing the uncontrolled water from the cylinder. When closure is initiated, or there is a failure of the oil supply, the valve is closed by the uncontrolled water pressure.

Sphere valves are manufactured in sizes up to 3.5 m diameter and working pressures up to 500 m. Smaller size valves up to 2 m diameter are available up to working pressures of 1000 m head.

3.9. Matching terminal-discharge valves and guard valves

Figure 3.18 shows a typical arrangement of bottom-outlet valves where a hollow-cone valve is backed by a butterfly valve. The diameters of the two valves have to be matched so that the hollow-cone valve is smaller than the butterfly valve. The change in the diameter of the conduit is effected by a taper section between the valves. This produces a back-pressure and prevents cavitation in the pipe and the butterfly valve when the hollow-cone valve is in the fully open position. In order to calculate the difference in the size of the valves, the loss coefficient of the hollow-cone valve between 70% to full open must be known. As a very approximate guide, the area of the terminal section should be about 60% of that of the section where the butterfly valve is located.

Operationally the butterfly valve is used only for emergency closure; otherwise the valve is actuated under balanced conditions. To enable this to be carried out, water under reservoir head is admitted to both sections of pipe upstream and downstream of the butterfly valve. If the section of pipe upstream of the valve remains under pressure, admission of water under reservoir head is required only to the downstream side. Apart from means of draining the sections of pipe, provision for releasing air must be provided. The closure characteristic of the butterfly valve must be such as to minimise water hammer.

The matching of a terminal-discharge valve with a guard valve can also apply to other combinations of valves apart from that illustrated in Figure 3.18. It is likely to be more critical when a hollow-cone valve is used for discharge because of the low loss characteristic of a fully open hollow-cone valve.

Appendix 3.1. Main applications, advantages and disadvantages of various types of hydraulic valve

Type	Main Application	Advantages	Disadvantages
1. Sluice valves	Controlling flow at low velocities. Closure and opening of flow.	Low cost. Simple. Reliable.	Unsupported valve blade during raising and lowering. Eddy shedding from blade tip.
2. Butterfly valves	Closure device in pressure conduit.	Relatively low loss coefficient. Available in large sizes. Capable of working at high-heads. Closure by gravity can be arranged.	Normally opened under balanced conditions. Possibility of blade flutter. Possibility of eddy shedding from blade tips.
3. Hollow-cone valves (Howell-Bunger valves)	Terminal-discharge.	Very efficient energy-dissipation valve. Simple construction. Relatively low cost. Can be operated electro-mechanically or by oil hydraulics. Good discharge coefficient. Available in large sizes. Least flow obstruction of any terminal-discharge valve.	Seal of sliding sleeve may leak. Can trap debris.
4. Hollow-jet valves	Terminal-discharge	Dissipates energy. Can be arranged to discharge into a stilling basin at an angle.	Less efficient energy dissipator than 4. Lower coefficient of discharge than 4. Greater cost than 4. Fluid passages can get blocked. Internal moving parts. Inspection and servicing requires removal of valve.
5. Needle valves	Terminal-discharge.	Dissipates energy. Can be used as an in-line pressure-reducing valve	Less efficient energy dissipator than 4. Greater cost than 4. Low coefficient of discharge. Fluid passages can get blocked. Internal moving parts. Inspection and servicing requires removal of valve.

Type	Main Application	Advantages	Disadvantages
6. Pressure-reducing valves, perforated cylinder type	Pressure control in closed pipes.	Pressure control.	Orifices in perforated cylinder can be blocked by debris. Internal moving parts. Inspection and servicing requires removal of valve.
7. Sphere valves/ Rotary valves	Shut-off control in high pressure conduits.	Low loss coefficient. Shuts drop tight. Manufactured in large sizes. Capable of working at high-heads. Can be supplied with a maintenance seal.	Greater cost than 2.

References

1. Bramham, H T (1979): Developments in Through Flow Butterfly Valves, *Water Power and Dam Construction*, Mar.
2. Ellis, J; Mualla, W (1984): Dynamic Behaviour of Safety Butterfly Valves, *Water Power and Dam Construction*, Apr, p.26–81
3. Miller, D S (1978): *Internal Flow Systems*, B.H.R.A., Fluid Engineering.
4. Rahmeyer, W J (1981): Cavitation Limits for Valves, *Journ. A.W.W.A.*, Nov, p.582–584.
5. Instrument Society America: *Control Valve Capacity Test Procedure*, ISA–S39.2, Pittsburgh, Pa.
6. Mercer, A G (1970): Vane Failures of Hollow-cone Valves, *I.A.H.R. Symposium*, Stockholm, paper G4.
7. Nielson, F M; Pickett, E B (1979): Corps of Engineers Experiences with Flow-induced Vibrations, *19th I.A.H.R. Congress, Karlsruhe*, paper C3.
8. Falvey, H T (1979): Bureau of Reclamation Experience with Flow-induced Vibrations, *19th I.A.H.R. Congress, Karlsruhe*, paper C2.
9. Rao, P V; Patel, G G (1985): Hydraulic Characteristics of Cone Valves with Different Angles, *Irrigation and Power*, Jul, p.233–243.
10. Brighouse, B A; Chang, E (1982): *Design Data on Deflector Hoods for Hollow-cone Outlet Valves*, B.H.R.A., Report 1939, Dec.
11. Brighouse, B A; Chang, E (1982): *Monasavu Hydroelectric Scheme, Fiji, Part 2, Model Study of Howell-Bunger Valve for the Controlled Filling Outlet*, B.H.R.A., Report RR1818, Mar.

4
Trashracks, screens and debris

4.1. Trashracks and screens in submerged intakes

The terms trashrack and screen are synonymous, the former being favoured in the US and the latter in the UK. They are provided at intakes to turbine or power penstocks, and at pumping stations, to prevent debris in irrigation canals and where valves liable to blockage are installed. They are either of the mobile type, withdrawable to the surface for cleaning, or form a fixed installation cleaned by raking from an overhead gantry.

Figure 4.1 shows a screen mounted on a rail wheel bogey protecting an abutment intake on a rock-fill dam. This installation is at the Kotmale dam in Sri Lanka. A similar arrangement was used at the Mangla dam in Pakistan. The screen carriage moves on rails and is hoisted clear of the water for cleaning. The semicircular shape of the screen increases the free area.

Figure 4.2 illustrates removable screen installations used principally to protect intakes at concrete dams. If it is necessary to provide a hood over the intake to suppress vorticity a rectangular, withdrawable screen is used. A problem can then arise if a long section of debris becomes stuck in the screen and the screen cannot be withdrawn.

There are at least four criteria which must be considered when designing screens:

- Differential design head, that is, head loss across the screen plus head loss due to the accumulation of debris. (In submerged screens a differential head of 3m is often used.)
- Spacing of screen bars. (This depends on the capacity of turbines or pumps to pass solid objects. A frequent spacing is 100 mm, although 75 mm or less is sometimes used.)
- Head loss across the screen
- Vibration

Failures of screens due to vibration of the screen bars have been recorded[1,2,4]. Vibration depends on the natural frequency of the screen bars,

Figure 4.1. Screen protecting an abutment intake screen hoisted to surface for cleaning

the forcing frequency and the possibility of resonance developing[3]. Vibration occurs when the two frequencies approach resonance.

The natural frequency of oscillation of screen bars in water is given by:

$$f_n = \frac{\alpha}{2\pi} \sqrt{\left(\frac{EIg}{(m+m_w)L^3} \right)}$$

where f_n = natural frequency
E = Young's modulus
I = moment of inertia of screen bar
m = mass of screen bar
m_w = added mass of water; that is the mass of water vibrating with the bar
L = length of bar between supports
g = gravitational constant
α = a coefficient depending on the end fixity of the bar (bars are normally welded to the supporting grid frame, and α is between 16 to 20 for bars between 60 to 70 mm deep, having a thickness to depth ratio of 5:1)

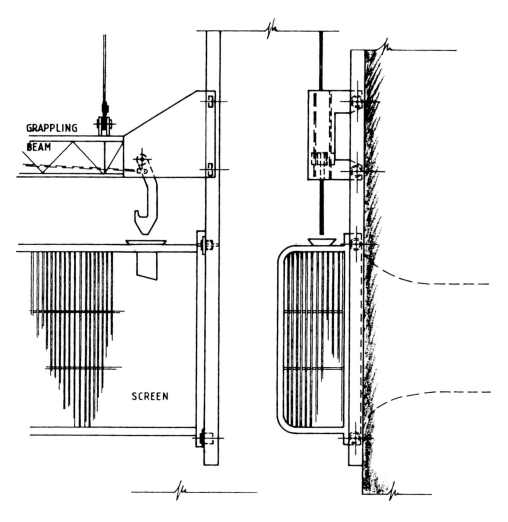

PART FRONT ELEVATION   SIDE ELEVATION

Figure 4.2. Removable screen installation

From Levin,[1,2] m_w can be approximated:

m_w = equivalent mass of water of the same volume as screen bar $X(b/d)$

where b = the effective spacing between bars
       d = the thickness of the bar
or

$$m_w = \frac{m}{8} \times \frac{b}{d}$$

The work of Levin[2] suggests that the value of the b term be limited to 0.55 times bar depth for a bar with a depth to thickness ratio 10 and to 1.0 times bar depth for a bar with depth to thickness ratio 5. In practice, the effective spacing between bars will be greater than the bar depth but the computed value of *b* should be based on the suggested relationship. This will yield conservative results.

The forcing frequency due to vortex shedding at the downstream edge of the screen bar is given by:

$$f_f = SV/d$$

where f_f = forcing frequency
       S = Strouhal number
       V = approach velocity
       d = thickness of screen bar

The Strouhal number depends on the spacing between bars and the shape of the bars. Levin[1,2] gives detailed information. For most design purposes the limit value of the Strouhal number applies when the bar spacing to bar thickness number is 5 or greater. For a bar fully rounded upstream and downstream the limit value of the Strouhal number is 0.265, and for a screen bar with sharp rectangular profile the number is 0.155.

The problem in selecting velocity is the difference between the flow across the bars when the screen has been cleaned and when it is partially blinded by debris. It is suggested that a range of values be used, at the lower end the nett velocity through the bars and at the upper end a value 3 times greater. Although screens are not normally operated so that debris is permitted to accumulate to this extent, the local velocity of a partially-blinded screen may significantly exceed the average velocity. It is recommended that the natural frequency of oscillation of the screen bars should be 2 to 2.5 times apart from the forcing frequency. In the diagram of the velocity profile at La Plate Taille[4], the variation of flow velocities about the mean value appears to lie between 0.5 and 1.8.

4.2. Trashracks and screens in culverts and river courses

Most screen installations are at the entry to a culvert in order that the risk of blockage occurs outside the culvert and where debris can be more easily removed. A secondary purpose is to prevent unauthorised entry into the culvert and for this reason a safety screen is usually also provided downstream of a culvert. Magenis[5] reports an extraordinarily high number of local councils (91%) and water authorities (95%) in the UK who have experienced serious problems with screens. Structural failures were significant, 14% for local councils and 18% for water authorities.

The survey highlighted the inability of many screens to satisfy basic criteria:

- To pass the maximum of flow when partially blocked to match the capacity of the culvert protected.
- To allow safe clearance of debris under normal and adverse conditions.
- To prevent all debris which would cause a blockage from entering the culvert.
- To remain structurally sound under all conditions.

To satisfy the first criterion and take into account the possibility of a screen becoming completely blocked a bypass should be provided. In urban river courses it may present a physical problem to accommodate a bypass.

4.3. Screen instrumentation

Screen instrumentation for submerged inlets

It is a usual requirement that the head loss across the screen be measured to give an indication when the screen has to be cleaned. At depth of inlet of 30 m or more the most suitable instrument is the bubbler device, because it can measure differential head by means of a sensitive bridge.

Pressure transducers located at the approach and downstream of a screen located at depth have to be limited in the range of head to be measured. The accuracy of the instrument is increased if the range of operational head is limited, irrespective of depth of installation. For instance, if pressure transducers with an accuracy of ±0.75% are installed at 30 m depth and measure total head, the reading can be 0.45 m in error.

Bubbler devices are considerably more expensive to install than pressure transducers and require maintenance. To overcome this, at least one installation uses tubes which are brought to the surface and the water level is measured by tank gauges. The output from the tank gauges is compared and the difference is displayed.

Screen instrumentation for screens in free surface water

Pressure transducers or ultrasonic level-measuring instruments are used and are positioned one instrument upstream and another downstream of the screen. The instruments should be located in a stilling well. The output from the instruments can be displayed separately or integrated to show differential head. It is a frequent practice for staged warnings to be sounded as the head loss across the screen builds up. A difficulty sometimes experienced in small river courses is the rapid build up of head loss which can be caused by a single large object such as a tree.

4.4. Screen-raking

A variety of screen-raking machinery for fixed-screens is available. Figure 4.3 illustrates two types. Screens which are hoisted to the surface (such as the one shown in Figure 4.1) have to be raked by hand using special purpose

Figure 4.3. Screen raking machinery

combs. An auxiliary crane is provided at the cleaning platform for this screen to handle logs or trees which have been caught up in the screen bars.

4.5. Debris

In underflow gates, debris will not normally be discharged until a gate is 70% to 80% open. Under conditions of drowned discharge, debris will become trapped in the hydraulic jump which forms in the stilling basin, and may recirculate for an appreciable time. Floating oil cans or other metal containers which repeatedly impact the submerged sections of gate arms or structural stiffeners of the skinplate can be a noise nuisance at gate installations close to dwellings. Floating debris can also cause damage to gate equipment and to paintwork on gates.

At overflow gates or overflow sections of gates, debris and flotsam will be discharged from upstream but can build up at flow breakers where these have been provided to vent the nappe or become trapped by the discharge roller which forms downstream of the gate (Figure 2.24).

In bottom-hinged flap gates which recess into the river bed, floating timber trapped upstream of the gate discharge roller can cause operational problems.

At flood diversion channels which are controlled by gates, and which incorporate a fixed weir alongside, a boom is often fixed across the flood channel to divert floating debris to the weir.

A floating boom is often placed at the exit of a stilling basin to prevent debris discharged over the weir from refluxing and entering the stilling basin when the flood relief gates are shut. It can then remain trapped in the stilling basin when discharge under the gates commences.

Floating booms for diverting debris are positioned across a waterway at an angle of 30° to 45° to assist flotsam to be driven towards the bank for clearance. Slack must be provided in the stringer cable to allow for a rise in water level during a flood.

In rivers which carry a large amount of flotsam during the flood season, an appreciable load can accumulate on structural stiffeners of gates. To prevent this, the rear of the gate skinplate is sometimes protected by wire mesh.

References

1. Levin, L (1967): Problèmes de Perte de Charge et de Stabilité des Grilles de Prise d'Eau, *La Huille Blanche*, Vol. 22, No.3, p 271–278.
2. Levin, L (1967): Etude Hydraulique des Grilles de Prise d'Eau, *Proc. 7th Gen. Meeting, I.A.H.R. Lisbon*, Vol. 1, p C11.
3. Sell, L E (1971): Hydroelectric Power Plant Trashrack Design, *Proc. A.S.C.E., J Power Div*, Vol. 97, Jan, No.PO1.
4. Vanbellingen, R; Lejeune, A; Marchal, J; Poels, M; Salhoul, M (1982): Vibration of Screen at La Plate Taille Hydro Storage Power Station in Belgium, *Int. Conference on Flow Induced Vibrations in Fluid Engineering*, Reading, England, Sep, p B2.
5. Magenis, S E (1988): *Trash Screens in Urban Areas*, I.W.E.M., River Eng. Section, Jan.

5
Structural considerations

The design of gates and valves can be analysed by conventional structural or stress analyses, if the curvature of the skin plate of radial gates is ignored. National codes of structural design are used, but load factors are introduced to reduce working stresses by 15 to 25%. The extent and magnitude of load factors applied is usually varied for different design elements, operating conditions and the possible occurrence of extreme events which cannot be reliably quantified. Examples are ship or vessel impact, increases in weight due to entrained water or floating debris, exceptional tide levels, jammed foreign bodies, increased loads due to impeded movement caused by solid freezing, ice impact and possibly irregular settlement and deformation of the foundation works. Some designers or specification writers require that a corrosion allowance be added after calculations have been carried out determining the sizes of members and plates. A reduction in working stresses is a more rational approach.

The stress analysis specific to gates, which is not normally encountered in onshore structures, is the combination of panel and bending stresses in stiffened plates subject to hydraulic pressure. The stiffening sections welded to a skin plate of a gate (Figure 5.1) act compositely with the skin plate to form the top flange of a beam. The composite beam is, in bending, transferring the hydrostatic load to other beams or support members. The skin plate is also subject to a panel stress at right angles to the bending stress of the beam. These stresses have to be combined.

Stresses in plates are dealt with in Timoshenko and Woinowsky-Krieger[1] and are tabulated in convenient form in Roark and Young[2]. The two cases most frequently met in gate analyses are reproduced in Table 5.1 (a) and (b). At centre of long edge, max. stress

$$\sigma = \frac{-\beta_1 q b^2}{t^2}$$

where q = intensity of load (hydrostatic pressure)
       t = plate thickness

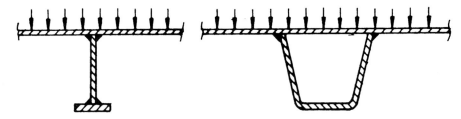

Figure 5.1. Stiffening members welded to a plate subject to hydrostatic load

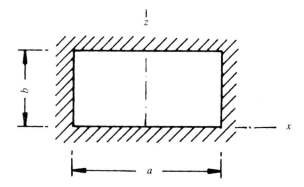

Illustration of a rectangular plate, all edges fixed, uniform load over entire plate

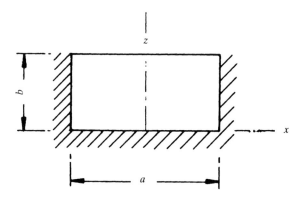

Illustration of a rectangular plate, three edges fixes, one edge (a) free, uniform load over entire plate

Table 5.1. (a) Variables in stress and deflection equations of plates (all edges fixed)

a/b	1.0	1.2	1.4	1.6	1.8	2.0	∞
β_1	0.3078	0.3834	0.4356	0.4680	0.4872	0.4974	0.5000
β_2	0.1386	0.1794	0.2094	0.2286	0.2406	0.2472	0.2500
α	0.0138	0.0188	0.0226	0.0251	0.0267	0.0277	0.0284

At the centre of the plate

$$\sigma = \frac{\beta_2 qb^2}{t^2}$$

and the maximum deflection at the centre of the plate

$$y = \frac{\alpha qb^4}{Et^3}$$

where E = Young's modulus

$$\text{At } x = 0, z = 0, \text{ max. stress } \sigma_b = \frac{-\beta_1 qb^2}{t_2} \text{ and } R = \gamma_1 qb$$

where R = the reaction force normal to the plate surface exerted by the boundary support on the edge of the plate in N/mm

The units of q are N/mm^2 and b is in mm

$$\text{At } x = 0, z = b \quad \sigma_a = \frac{\beta_2 qb^2}{t^2}$$

$$\text{At } x = \pm(a/2), z = b \quad \sigma_a = \frac{-\beta_3 qb^2}{t^2} \text{ and } R = \gamma_2 qb$$

Stiffener beams are considered to be continuous. Figure 5.2 shows the loading condition, the bending moment diagram and the associated skin plate which forms the upper flange of the beam. The reduction factors V_1 and V_2 depend on the ratio of the panel support dimensions L and B and are listed in Table 5.2. This computation of the width of the panel acting

Table 5.1. (b) Variables in stress equations of plates (three edges fixed)

a/b	0.25	0.50	0.75	1.0	1.5	2.0	3.0
β_1	0.020	0.081	0.173	0.321	0.727	1.226	2.105
β_2	0.016	0.066	0.148	0.259	0.484	0.605	0.519
β_3	0.031	0.126	0.286	0.511	1.073	1.568	1.982
γ_1	0.114	0.230	0.341	0.457	0.673	0.845	1.012
γ_2	0.125	0.248	0.371	0.510	0.859	1.212	1.627

compositely with the stiffener section to form a beam is the methodology given in DIN 19704[3]. This standard also gives guidance on many other aspects of design. It is under revision (1993) so that it is based on load factors and not on safe working stresses.

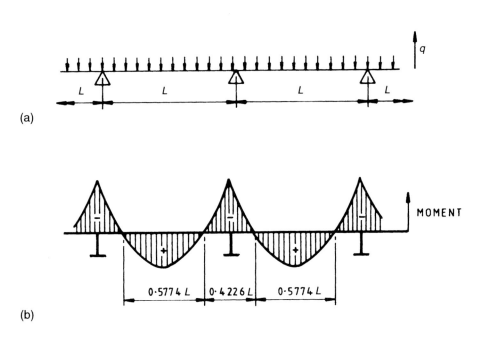

(a)

(b)

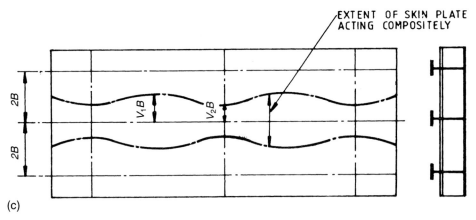

(c)

Figure 5.2. Composite action of stiffeners and beams with skin plate: (a) Load diagram; (b) Bending moment diagram; (c) Skin plate acting compositely in bending with stiffeners

Table 5.2 Reduction factors V_1 and V_2 for skin plate acting compositely in bending with stiffeners (Poisson's ratio 0.3)

L/B	V_1	V_2
∞	1	1
20	0.984	0.861
10	0.938	0.753
6.67	0.867	0.660
5	0.783	0.580
4	0.697	0.512
3.33	0.616	0.453
2.86	0.545	0.404
2.5	0.484	0.363
2.22	0.433	0.324
2	0.391	0.295
1.67	0.325	0.250
1.43	0.276	0.215
1.25	0.241	0.190
1	0.195	0.155

A method and data for analysing curved skin plates and stiffener beams for radial gates are given in Wickert and Smausser[4].

The biaxial stresses represented by the panel stress acting at right angles to the bending stress of the beam are combined so as to calculate the equivalent stress σ_e, where:

$$\sigma_e = \sqrt{(\sigma_x^2 + \sigma_y^2 - \sigma_x\sigma_y + 3\tau^2)}$$

σ_x and σ_y are the normal stresses in orthogonal directions, that is the panel stress and the beam bending stress. They are substituted with their signs. τ is the shear stress, which is calculated as:

$$\tau = TS/Id \text{ or for a member with I or box section as } \tau = T/A$$

where T = shear force
       A = cross-sectional area of web plate
       S = static moment about the centroid of the section of part of the cross-section between the point concerned and the extreme fibres
       I = moment of inertia
       d = web plate thickness

References

1. Timoshenko, S P; Woinowsky-Krieger, S (1970): *Theory of Plates and Shells*, 2nd Edition, McGraw-Hill.
2. Roark, R J; Young, W C (1975): *Formulas for Stress and Strain*, 5th Edition, McGraw-Hill.
3. DIN 19704 (1976): *Hydraulic Steel Structures—Criteria for Design and Calculation*.
4. Wickert, G; *Schmausser*, G V (1971): Stahlwasserbau, Springer Verlag, Berlin-Heidelberg-New York.

6
Operating machinery

Operating machinery for gates is either by electro-mechanical drives or by oil hydraulics. In some installations screw jacks have been used as shown in Figure 6.2.

Electro-mechanical drives consist of electric motors driving hoisting drums or sprockets through multi-stage reduction gear boxes. The gates are raised by ropes or chains. The gates close by gravity. In electro-mechanical drives gravity cannot be used for emergency closure in the event of failure of the electricity supply because the reverse efficiency of gear trains with reduction of the order of 2000:1 or greater is self-sustaining.

In most gates, the critical emergency operation is opening. To effect this under gravity the gate has to be counterbalanced so that the closure motion requires the main drive effort, that is, the mass of the gate is overbalanced. The gear reduction required to effect closure can then be kept low to ensure that the gear train does not become self-sustaining when the gate is raised. While gates have been designed to operate on a 'fail safe' basis, this does not appear to have been done with gates with electro-mechanical drives.

Oil hydraulics applied to a hoisting cylinder, usually referred to as a servo-motor, can actuate closure as well as opening of a gate. In many cases the two drive systems are interchangeable, and the choice may depend on the layout or be influenced by visual considerations at locations where the appearance of electro-mechanical machinery above the abutments or piers is not acceptable. Another factor may be possible pollution of a watercourse in the event of failure of a hydraulic pipe or of a flexible hose. This can be overcome by using an environmentally compatible hydraulic fluid, or locating the cylinder and pipework in a separate chamber as shown in Figure 2.25.

Gate hoisting by ropes or chains introduces an elastic suspension. Therefore certain types of gate have to be close coupled to their servo-motor or be rigidly connected. This applies to control gates in submerged outlets and is dealt with in other chapters.

In Chapter 2 it was mentioned that some bottom-hinged flap gates must also be close coupled to their actuators, most frequently one or two hydraulic cylinders. This arises in tidal barrages where the tide level can exceed the pond or reach level and reverse the normal thrust on the cylinder. A similar consideration applies to fish-belly gates where the fish-belly section is sealed and becomes buoyant when submerged. The gate then exerts an upthrust which the cylinder must resist.

6.1. Electro-mechanical drives

Single motor drives with line shafting driving multi-reduction gearboxes at each end are a common method of layout. Apart from simplicity of control, it is possible to mount two motors to duplicate the drive in the event of failure of one motor and to arrange for manual winding of the motor extension shaft on mains failure. Figure 6.1 shows two frequent layouts of single motor drive. Two hoisting ropes per side are frequently used. They are connected at the gate anchorage point by a compensating beam to allow for differential rope stretch. The articulation of the compensating beam is restricted so that, in the event of failure of one rope, hoisting can be continued with the other one, Figure 6.3.

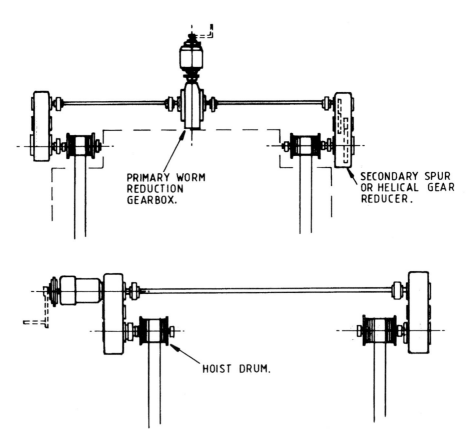

PRIMARY WORM
REDUCTION
GEARBOX.

SECONDARY SPUR
OR HELICAL GEAR
REDUCER.

HOIST DRUM.

Figure 6.1 Arrangements of single motor drive of radial gates

Similar layouts are used when chains are the means of gate suspension. In that case, the hoist drums are replaced by chain sprockets.

If squirrel-cage induction motors are used, they should be of the high torque or double-cage design. Star delta starting is not suitable for hoist motors, even with Wauchop 'no break' winding. A standard squirrel-cage motor will develop about 150% full-load torque on starting, whereas 200% is characteristic of a high torque motor. With star delta winding the starting torque is only 54% of the motor full-load torque, and in order to start and accelerate a hoisting load the motor has to be substantially oversized. This in turn creates design problems, since in running up to full speed a standard squirrel-cage motor will develop about 280% of full-load torque at 65% of its rated speed. Similar considerations apply to auto-transformer starting, although the choice of tap gives greater flexibility in the starting characteristics.

In a wide gate or where a clear, unobstructed watercourse is required, independent drive at each side may be necessary. Up to 30 kW per motor size, electrical cross-synchronisation can be provided by means of power Selsyns. These are three-phase AC motors coupled to the drive motors. The primary and secondary windings of the power Selsyns are interconnected and will transmit full synchronising torque at all speeds (Figure 6.2). The development in the last ten years of Selsyn stabilisers to synchronise the receiver Selsyn with the transmitter has overcome one of the limitations of the system which was associated with the requirements to synchronise the machines at standstill and the violent standstill synchronisation which could occur.

It is possible to design such a system so that one motor can lift a gate, in the event of a breakdown of the other machine, by transferring the power required through the Selsyn machines. This necessitates oversizing of the motors and the Selsyns. If this is adopted, the start-up characteristics have to be calculated with precision and the mechanical components designed accordingly. Otherwise if the pull-out torque is developed by the normal operation of the hoist motors, shafts, keyways, couplings and gears could be sheared or damaged. It is not possible to transfer manual winding at one motor to the other in the event of a mains failure, since the primary windings of the power Selsyns have to be energised from the mains for synchronisation to be effective.

Alternative linked drives can be achieved by synchronous motors fed from a variable-frequency supply or thyristor-controlled DC motors with servo controls. The former system also offers the benefit of speed variation, if it should be required. The starting characteristics are not as good as can be obtained with DC motors or high torque squirrel-cage AC motors. Thyristor-controlled DC motors with servo controls can be up to 55 kW per motor, or even larger.

With any thyristor control, great care should be taken with signal cables, particularly those transmitting signals from solid-state devices. Signal cables should be screened and not run alongside motor feeder or control cables. In some circumstances a clean supply should be considered.

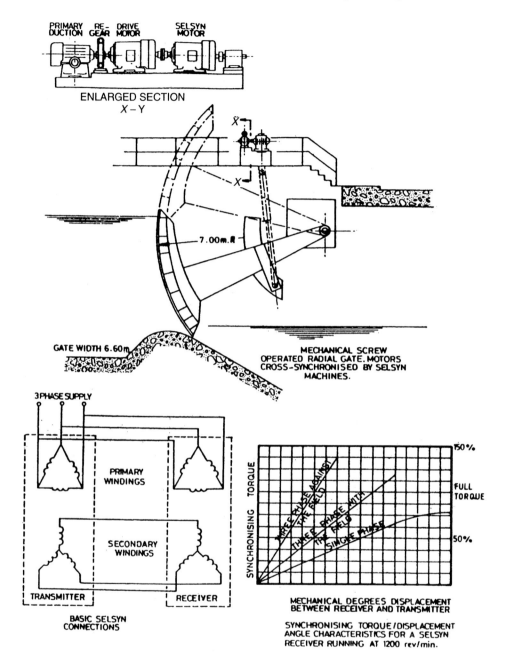

Figure 6.2. Principles and application of a power Selsyn motor drive

The hoist ropes can be anchored to the gate either upstream of the skin
plate (Figure 6.3(a)) or downstream (Figure 6.3(b)). Upstream anchorage
with the ropes in contact with a wear plate welded to the skin plate has the

disadvantage that debris can become lodged between the ropes and the skin plate and that the ropes are immersed during the majority of their working life. In spite of these disadvantages it is sometimes used because the layout of the hoist machinery shown in Figure 6.1 is often difficult to achieve or is incompatible with the location of the gear box drive shaft upstream of the gate. However, it should be avoided if at all possible.

A typical arrangement of hoist machinery for vertical-lift gates is shown in Figure 6.4.

6.2. Oil hydraulic operation

Oil hydraulic operation by a cylinder at either side of a gate permits more

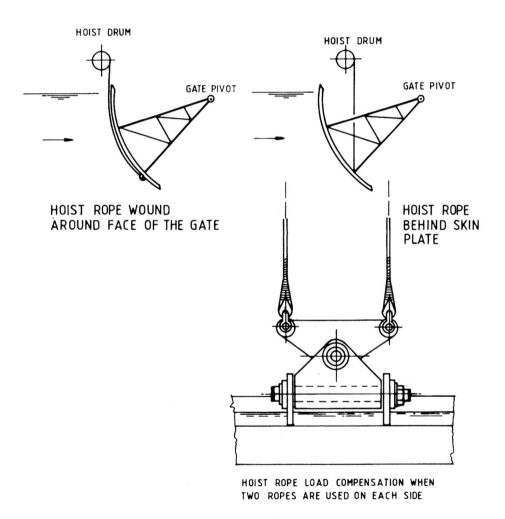

Figure 6.3. Hoist rope attachment to radial gates

Figure 6.4. Hoist machinery layout for an intake gate (Kotmale Dam—Sri Lanka)

flexibility in the layout of the operation machinery. It can also provide a high degree of redundancy in the power source when several gates are actuated in this way. In radial gates with an overflow flap it simplifies the hoisting arrangements.

Long hydraulic cylinders present problems of deflection of the piston rod, due to the mass of the cylinder assembly. This can result in rapid seal wear at the stuffing box.

The maximum extension of the piston may have to be limited to prevent flexing of the piston rod (Figure 6.5) to reduce wear at the stuffing box and deflection of the piston rod. Some designers favour short, large cylinders actuating radial gates at the gate arms. Another way is to mount the cylinders at or near their mid-point (Figure 6.6). Under earthquake conditions, when the piston is extended, a whipping motion can be set up leading to permanent set of the piston rod. Larger diameter shorter cylinders and the layout shown in Figure 6.6 can overcome this.

Radial- or vertical-lift gates are often operated by a cylinder on each side. It is possible to synchronise the movement of the cylinder by a loop servo system. However it requires a significantly more complex control circuit. In some cases this is not considered warranted and in radial and tilting gates of the fish-belly type the torsional stiffness of the gate is sometimes used to compensate for a degree of out of balance force.

Slide gates and roller gates, when controlling flow or opening and closing

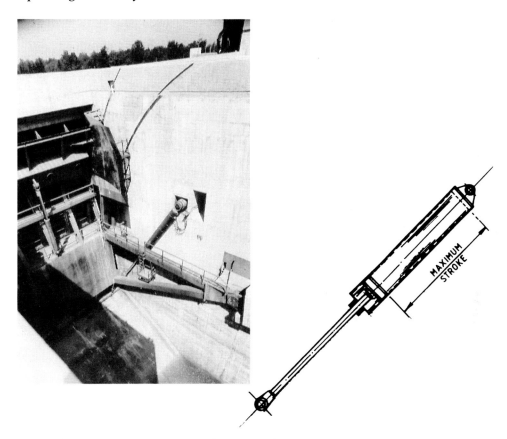

Figure 6.5. Maximum extention of piston rod to prevent flexing

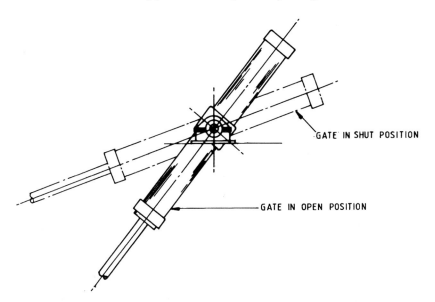

Figure 6.6. Gimbal mounting of hoist cylinder

under unbalanced hydraulic conditions, must be operated by a hydraulic servo-motor. Fluctuating and variable hydrodynamic forces act on the gate. To prevent these from causing gate vibration, the piston and the annulus side of the servo-motor are pressurised during opening and closing of the gate. In gates subject to high-heads of water and therefore high flow velocities, the closure speed must be controlled to prevent hydraulic downpull forces from accelerating the movement.

Operating gates are required to hold partial openings for a long time and the extent of the opening is often critical. Servo-motors are required to maintain pressure in the piston and annulus side of the piston during partial opening, although the hydraulic pumps are not in operation. This is effected by a hydraulic accumulator. When the pressure in the accumulator falls below a preset value, the pump or pumps are started up automatically.

In hydraulic cylinders leakage of oil occurs past the piston seals and increases as seals become worn. Therefore gates operated by hydraulic cylinders and in the open or partially open position, will gradually close. Gates are fitted with position indicators and these are utilised to signal a predetermined gate movement and to initiate a signal to restore gate position.

Features which should be part of a reliable oil hydraulic system consist of:

* Inlet strainers for the tank.
* Suction and delivery filters with 5 micron apertures.
* Two manually operated pumps.
* Directional control valves which can be manually operated in addition to electrical actuation by solenoids. If several valves have to be actuated manually in an emergency, they should be so grouped that they can be operated by one man and that they should all move in the same direction for the emergency condition.
* An off-loader valve which can be combined with a relief valve. The off-loader valve will prevent heat build up when the pump is running but is not delivering oil to a piston.
* Pipework of stainless steel fastened by saddles at close spacing, especially where there is a risk of earthquakes. Movement of the pipes due to seismic action can initiate vibration of high amplitude which can result in pipe fracture.
* Flexible hoses which are rated 50% above the system working pressure.
* Means of locking the oil at the piston raising port in the event of a pipe or flexible hose fracture. This can be effected by a pilot valve mounted on the cylinder which closes in the event of loss of system pressure.

6.3. Hoist speed

Hoist speeds of gates are conventionally 300 mm/min. In gates which control water level, hoisting is either carried out in steps followed by a dwell period, both of which are controlled by timers, or by set point control. The subject is further discussed in Chapter 11.

The final closure speed of servo-motors is usually decelerated. This can be effected by using hydraulically-cushioned cylinders. That is, pistons which have a stepped crown which mates with a recess in the cylinder cap. If variable delivery hydraulic pumps are used, the pump output is reduced when a limit switch is actuated shortly before the end of the stroke of the piston rod is reached. In dual pump operation, one pump is shut down by the limit switch.

For constant delivery by a hydraulic pump, the gate opening and closing speed will vary as the ratio of the cross-sectional area of the annulus and the piston side of the servo-motor. This is usually compensated by using two pumps with both pumps delivering only when the piston side is under full pressure. Two pumps also ensure greater reliability. In the event of failure of one pump, it is accepted that gate opening and closing speeds will vary.

7
Detail design aspects

7.1. Seals

Seals are required to prevent loss of water. While in many cases this may not be material, jets emitted at inadequately sealed sills, sides or soffits are a major source of gate vibration and gate noise. In addition, seal leakage in gates subjected to high-head can cause long-term damage to downstream concrete works. The selection of seals and the design of their mountings is therefore important.

Because the fluctuating pressure due to flow through gaps is a major cause of gate vibration, aspects of seal arrangements are also discussed in Chapter 10, Gate Vibration.

Modern seals are extruded or moulded from an elastomer. Block seals of timber or lignum vitae which were fitted to older installations are not suitable, as explained in Chapter 10. The usual materials are natural rubber or polychloroprene, commercially known as Neoprene. The elastomers are compounded to attain the required properties for seal application such as tensile strength, tear resistance, low water absorption, compression set and be ultraviolet resistant and contain anti-oxidants. Seals are normally specified to have a Shore A hardness of 65 with a tolerance of ±5. High-head gates often have seals of greater hardness. At hardness significantly lower than 65, the coefficient of friction between a seal and a stainless steel sliding surface increases. The coefficient is also affected by the surface finish of the seal contact face.

Approximate values are:

- Shore A 55 hardness, coefficient of friction 0.8
- Shore A 65 hardness, coefficient of friction 0.7
- Shore A 80 hardness, coefficient of friction 0.6

For PTFE covered seals this reduces to 0.1.

Side and top seals rely on the water pressure to aid sealing. The gap between a gate skin plate and the side rubbing strip must be sufficiently wide to permit inaccuracies of tracking and deflection under load, including those of the sluiceway walls. The seals must be able to accommodate these variations. If the gap is made too wide, the seal may not develop adequate contact pressure or extrude through the gap under the hydrostatic head. Seals should be preset, that is the stem should be placed under deflection, but not under compression, because the force required to compress a seal bulb is very much greater than that required to deflect the stem of a seal. If friction is to be kept to a minimum, a preset of 3–5 mm is advisable. Less will be adequate for effective sealing. Bottom seals rely on the weight of the gate to provide the contact pressure for sealing. Figure 7.1 shows typical seal shapes.

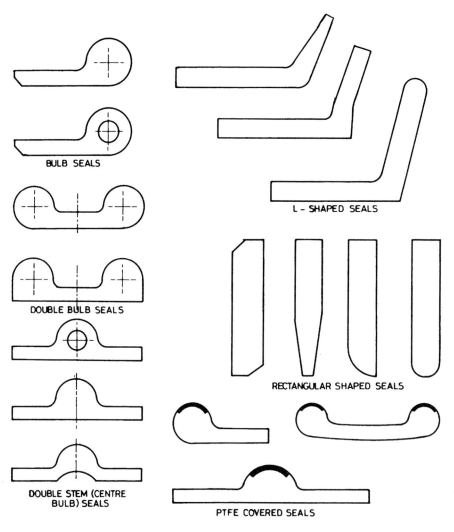

Figure 7.1. Typical seal shapes

Bulb seals in the shape of a musical note

These are more frequently used in the form of the solid rather than hollow bulb type. The solid-bulb seal has a reduced contact area because it deforms less than the hollow-bulb type under water pressure and is less liable to compression set.

Double-bulb seals

The main application is when sealing against a reversal of head is required as occurs in gates in a tidal river.
Types of gates shown in Figure 7.2:

a) Vertical-lift gate
b) Radial gate
c) Submergible radial gate
d) Bottom-hinged flap gate
e) Mitre gate
f)   Drum gate
g) Two-leaf vertical-lift gate
h) Vertical-lift gate with flap
i)   Bear-trap gate
k) Cylinder gate
l)   Vertical-lift tunnel gate
m) Culvert valve (reverse Tainter gate)

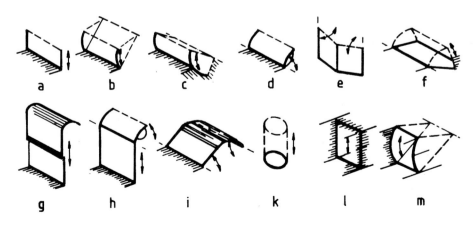

Figure 7.2. Gate boundaries requiring sealing

L-shaped seals

This type of seal can be more effective than the bulb seal, but can be used only for movement in one plane, that is in radial- or vertical-lift gates and not in bottom-hinged flap gates. The L-shaped seal is inherently more flexible and can blow through or 'fold under' if clearances become excessive.

Double-stem (centre-bulb) seals

This seal is used for face contact and particularly for sealing the top edge of submerged vertical-lift gates and radial gates in a culvert. These seals can be arranged so that they will move towards the seal contact plate under influence of the upstream head. At high-head gates when the seal is pressurised there is a risk that it could be extruded from its clamping. To prevent this a key is moulded at the end of the stems.

To ensure even clamping by seal retaining strips, the bolts passing through bulb and L-shaped seals should incorporate spacers.

Bolt holes through the skinplate for seal clamping are a frequent source of leakage. This is avoided by fitting nylon washers or washers faced with an elastomer under the nuts.

Splices in seals should be hot vulcanized, either in the factory or on site, and corners should be moulded. Special moulds are available for the junction of different seals.

When designing a gate sealing system it is desirable to arrange for sill and side seals to be in the same plane. This applies also to the uppermost or lintel seal for gates in conduit. It simplifies the sealing of the junction between the seals. If this cannot be achieved it is necessary to introduce a block seal to bridge the gap. These are difficult to engineer successfully.

Rubber has a tendency to contact bond when kept under high compression for a long time. Where gates are infrequently operated and are under high hydrostatic head, it may be advisable to uprate the coefficient of friction by 20%.

Radial gates: side sealing

The elastomeric seal bridges the gap between the skin plate and the sluice wall. On the pier or sluice wall a stainless steel contact plate provides the rubbing surface. Figure 7.3 (a) is a common arrangement. The gap between the weir plate and the seal contact face is of the order of 10 mm to permit minor inaccuracies in the width of the weir plate, the perpendicularity of the seal contact plate and the tracking of the gate. A greater gap could cause the seal to be extruded through the gap under the hydrostatic pressure of the upstream water.

Variations in tightening the bolts of the seal clamping plate can cause undulations of the seal. To ensure an even pressure a ferrule is inserted in

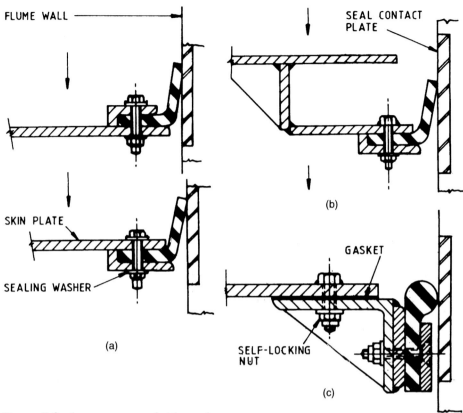

FLUME WALL

SEAL CONTACT PLATE

SKIN PLATE

SEALING WASHER

GASKET

(a)

(b)

SELF-LOCKING NUT

(c)

Figure 7.3. Arrangement of side seals

the holes in the seal, one millimetre less in thickness than the seal.

The arrangement of side sealing shown in Figure 7.3 (b) is used when the pivot centre and the origin of the weir plate radius do not coincide. This is sometimes adopted on large radial gates to reduce the hoisting force. In this case the seal mounting plate is on a radius whose origin is the pivot centre. This ensures that the seal is subjected only to radial sliding forces and not to lateral movement.

In Figure 7.3 (b) debris can lodge in the space between the weir plate and seal mounting plate. This can be a disadvantage. It is reduced by keeping the gap between the skin plate and the flume wall to a minimum.

Radial gates: sill sealing

The sill seal (Figure 7.4) is formed by a rectangular elastomeric section. The contact face can be square with the section, angled or rounded. Other seal profiles have been used but are incorrect for hydraulic reasons. This is more extensively discussed in Chapter 10 where it is pointed out that to prevent gate vibration there must be a sharp point of flow separation.

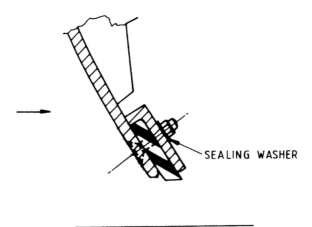

Figure 7.4. Arrangement of sill seal

The seal is set to deflect 2 to 3 mm when the gate seats on the sill. The seal should be arranged downstream of the weir plate. Upstream it causes flow separation. This favours the arrangement of the side seal downstream of the skin plate (as shown on the lower section of Figure 7.3 (a)). The seals are then in the same plane and a leakage path at the gate corners can be eliminated.

The sill seal can be used to take up only limited variation between the weir plate and the sill. Excessive projection of the seal below the weir plate causes it to deflect and leakage flow to occur. The angle at the sill should not be more acute than 45° otherwise lip deflection will result in difficulties to seal.

The sill seal can be abutted to the side seals in Figure 7.3 (a). Preferably the three seals are vulcanised together. When used in conjunction with any other side seal configuration such as Figure 7.3 (b) a leakage gap is created between the seals. To avoid this a block seal is introduced at the junction.

On large spillway gates, especially in tropical countries, there can be an appreciable temperature difference between the upstream face of the weir plate which is in contact with the water and the downstream face which may be exposed to the sun. The resulting curvature of the plate can create a leakage gap in the middle of the sill. Some gate manufacturers claim that by mounting the seal upstream of the weir plate, leakage can be avoided and that the effect of the turbulent hydraulic conditions created by mounting the seal in this manner can be overcome by stipulating that the gate is never opened by less than 100 mm. This avoids vibration of the gate due to the severe pressure fluctuations at low flows. Notwithstanding this consideration, the seal should not be placed on the upstream face. A criterion for minimum opening of the gate does not necessarily compensate for the turbulent hydraulic conditions caused by mounting the seal upstream of the weir plate.

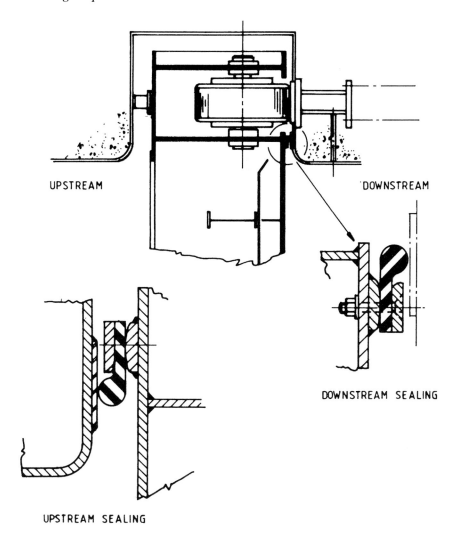

UPSTREAM

DOWNSTREAM

DOWNSTREAM SEALING

UPSTREAM SEALING

Figure 7.5. Side seal for a vertical lift gate

The sill seal and the side seal contact plates should be manufactured from stainless steel and this is also the practice for the side-guide roller contact plates.

Vertical-lift gates: side sealing

Side seals are usually of the musical note type, arranged as shown in Figure 7.5.

Vertical-lift gates: sill sealing

Sill seals (Figure 7.6) utilise the same rectangular elastomeric sections which are the practice on radial gates.

Vertical-lift gates: lintel sealing

For gates in conduit a double-stem, centre-bulb section is used to seal at the lintel. Using a seal of similar profile as the side seal and arranging it in the same plane, simplifies the detail of upper corners. Figure 7.7 shows the detail of a typical lintel seal. By admitting the upstream head of water to the underside of the seal, the pressure deflects the seal towards the contact plate. A similar seal arrangement is used at the lintel when radial gates are located in a conduit or operate as culvert valves.

Bottom-hinged flap or tilting gates: side sealing

Figure 7.8 shows different arrangements of side seals. Bottom-hinged flap gates are frequently used at tidal barrages because they can prevent the flow of estuarial salt water into the fresh water river course. In this application they may have to resist pressure in either direction. The seal arrangement of Figure 7.8 (a) will effect this. By using a seal with a single bulb the same layout can be used when sealing only against upstream head. This seal configuration is more flexible than that shown in Figure 7.8 (b) and can therefore accommodate a greater tolerance of the side staunching.

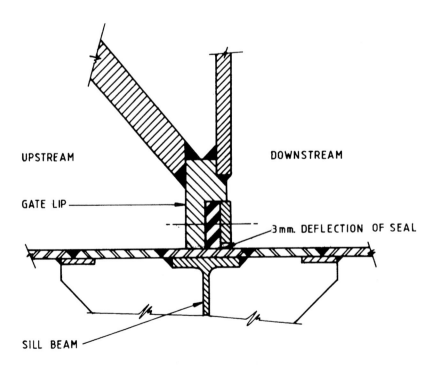

Figure 7.6. Sill seal for a high head for a vertical lift gate

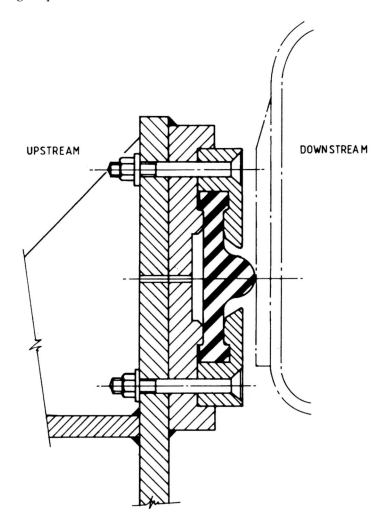

UPSTREAM   DOWNSTREAM

Figure 7.7. Lintel seal for a high head vertical lift gate

Bottom-hinged flap or tilting gates: hinge sealing

Figure 7.9 (a) shows the arrangement of a hinge seal to withstand only upstream head. Figure 7.9 (b) is capable of sealing both against upstream and downstream head. The seal shown in Figure 7.9 (b) can be used only when the gate pivots are located in the sluiceway walls. The seal contact face on the gate has to be accurately formed.

7.2. Guide and load rollers

Side-guide rollers for radial gates

Smaller gates of up to 40 m^2 of aspect area are sufficiently rigid not to require side-guide rollers. Larger gates are fitted with two guide rollers per side.

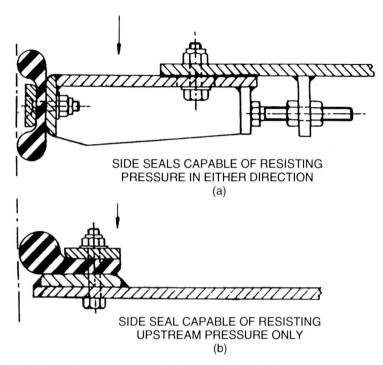

Figure 7.8. Side seal arrangements for bottom-hinged flap gates

Load rollers for vertical-lift gates

Load roller pressures are specified in DIN 19704[1] where clause 7.3.6 lists permissible Hertzian pressures.

In Table 7.1, σ_B is the lesser of the ultimate tensile strength of the two materials in contact with one another. For hardened contact faces, the stresses may be increased according to the hardness of the material.

The above values apply to rolling components temporarily immersed in water. For rollers permanently immersed in water and temporarily exposed to heavy water flows, the Hertz[1] stresses should be reduced:

for 0 to 300 revolutions under load per year by 10%
above 300 to 2 000 revolutions under load per year by 15%
above 2 000 to 20 000 revolutions under load per year by 30%
for more than 20 000 revolutions under load per year by 40%

For spherical revolving surfaces (crowned rolling face) with a diameter ratio of ≤ 15:1, the permissible Hertz[1] pressure between wheel and rail may be increased by 50%. The diameter ratio is the ratio of twice the radius of the crown of the load roller divided by the diameter of the load roller.

For Stoney rollers the values given in the table should be halved. This is due to uneven load distribution which occurs in Stoney roller trains. This does not apply to caterpillar rollers because they are not set in a fixed train.

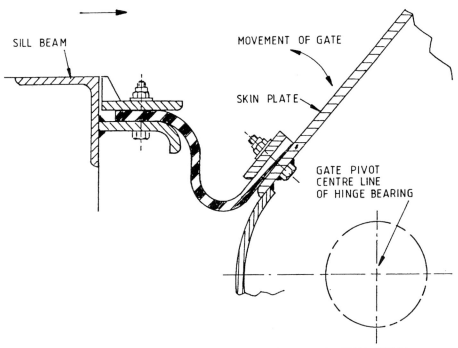

(a) SILL SEAL TO WITHSTAND UPSTREAM HEAD ONLY

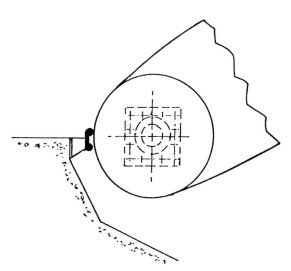

(b) SILL SEAL TO WITHSTAND UPSTREAM AND DOWNSTREAM HEAD

Figure 7.9. Hinge seal arrangements for bottom-hinged flap gates

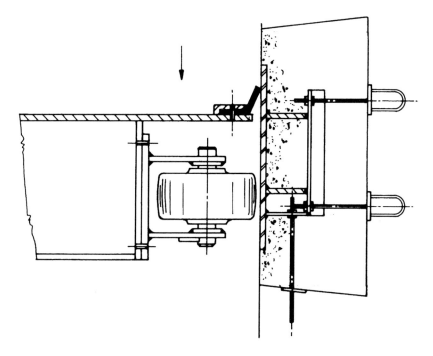

Figure 7.10. Side guide roller for radial gates

These criteria lead to very high stresses, particularly if manganese or nickel-manganese steels are used. US and UK practice is to use lower Hertzian pressures (0.7 to 0.8 of these values).

Bearings with either bronze alloy bushes or bushes with lubricant inserts as shown in Figure 7.15 are used because of their established performance in under water conditions. This applies whether the bearing is sealed or not, because the seal is liable to break down after 10 to 15 years due to ageing or wear. As mentioned in Chapter 2, graphite containing lubricants should not be used in conjunction with stainless steel as it causes electrolytic action which is accelerated under water.

If a plain bush is used the roller is often crowned to ensure that it will

Table 7.1. Permissible Hertz[1] Pressures σ_1 and σ_2

Operational condition	Contact surfaces	Gates (not frequently operated)	Gates and locks (frequently operated)
Rolling motions between non-hardened contact surfaces	Rail-wheel σ_1	1.85 σ_B cylindrical	1.6 σ_B cylindrical
	Roller-wheel or roller-axle σ_2	2.00 σ_B	1.9 σ_B

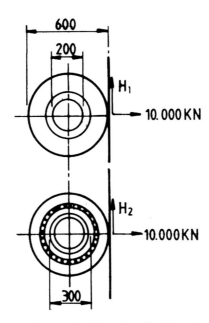

Figure 7.11. Load rollers with bronze bush and roller bearing

centralise and that the pressure distribution is symmetrical. On vertical-lift gates crowning of the load rollers compensates for deflection of the gate which otherwise would cause excessive contact pressure on one side of the roller. With self-aligning bearings of the angular contact type, the rollers have parallel faces and the articulation of the bearing will ensure centring of the roller.

The design of the sealing system is critical as is the backup to prevent ingress of water through any bolted face. The latter is achieved by 'O' rings. For maximum reliability it is necessary to ensure that the seals are lubricated and some gate manufacturers provide separate grease passages to each seal.

The difference in the coefficient of friction between a bronze bush and a roller bearing (0.1 as compared with 0.0018) makes an appreciable difference to the lifting force required, as the following example illustrates:

With a bronze bush:

$$H_1 = 10\,000 \times 0.1 \times \frac{200}{600} = 333 \text{ kN}$$

With roller bearing:

$$H_2 = 10\,000 \times 0.0018 \times \frac{300}{600} = 9 \text{ kN}$$

Where there are a number of rollers mounted on each side of a gate, adjustment to ensure even contact and appropriate load distribution is important and can be provided by making the shaft 'a' in Figure 7.12 eccentric. The roller shafts are rotated on assembly of the gate so that all the rollers are aligned accurately. When this has been carried out, the rollers are locked in position.

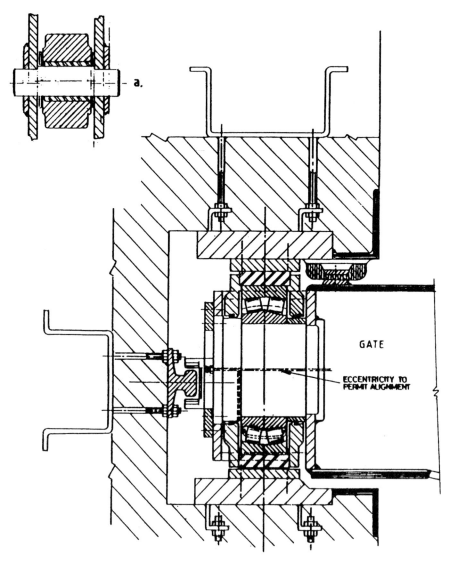

Figure 7.12. Roller assembly with eccentric shaft

7.3. Pivot bearings

Pivot bearings for trunnions of radial gates

Pivot bearings may be bushes (Figure 7.13) or incorporate self-aligning roller bearings (Figure 7.14). The coefficient of friction of a conventional lubricated bronze bearing material varies between 0.2 for starting conditions to 0.1–0.08 for running.

Self-lubricating bushes with insets of lubricating pads (Figure 7.15) have permissible bearing pressures of 200–300 bar, although in practice designers tend to limit bearing pressures to 70%–80% of the permissible values. Their starting coefficient of friction is stated to be 0.1 and under running conditions 0.08. Self-aligning double row roller bearings, such as the bearing shown in Figure 7.14, have a coefficient of friction of 0.0018.

Lining up the axis of two bearings and possible slight differences in settlement of the bearing supports can cause inaccuracies in tracking of the bearings, and in bushed bearings will therefore lead to non-uniform bearing pressure. The self-aligning bearing will of course compensate for this.

If two self-aligning bearings are incorporated in one bearing housing with the object of keeping the friction forces low they will not, of course, provide self-alignment of the assembled bearing.

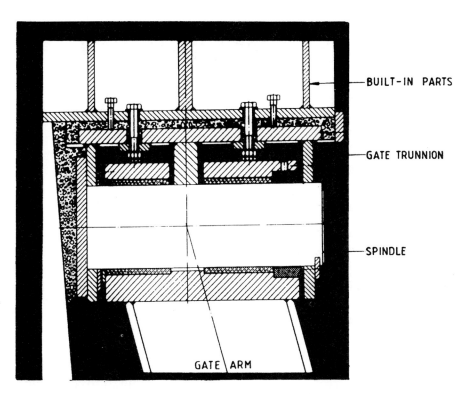

Figure 7.13. Gate trunnion with bushed bearing

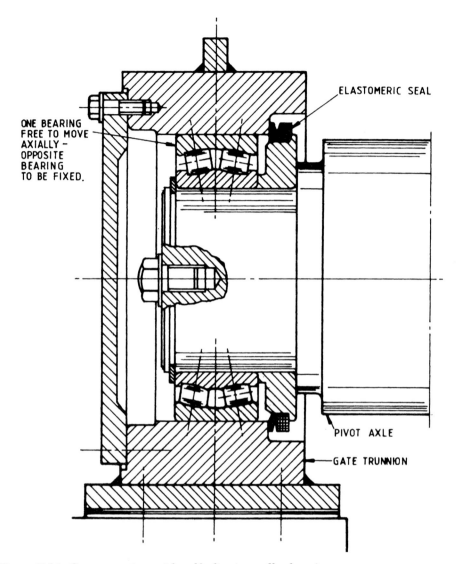

ELASTOMERIC SEAL

ONE BEARING
FREE TO MOVE
AXIALLY –
OPPOSITE
BEARING
TO BE FIXED.

PIVOT AXLE

GATE TRUNNION

Figure 7.14. Gate trunnion with self-aligning roller bearing

Spherical plain bearings (Figure 7.15) have been used for gate trunnions. They are substantially cheaper than the self-aligning, double-row roller bearings. The friction forces at the pivot bearings cause a transverse load on the pivot arms and must be taken into account in the stress analysis.

Bearing mounting of radial gates

A frequent arrangement is to recess the bearing mounting into the pier (Figure 7.15). For larger gates the reinforcement required to distribute the load and transfer it into the pier becomes complex.

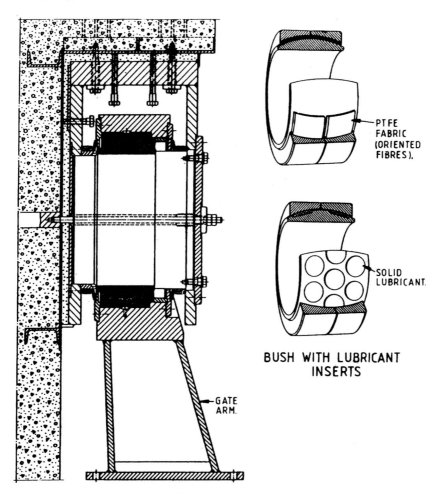

Figure 7.15. Gate trunnion with spherical plain bearing

When the load has to be transferred into the pier by prestressing cables, a pivot bearing mounting beam is used. Figure 7.16 shows a beam serving the anchorage of two gates. Prestressing cables for this application are encapsulated in oil to permit restressing after relaxation.

7.4. Limit switches

It is usual practice to provide backing-up limit switches to come into operation in the event of failure of an overwind or underwind limit switch. There is no practical way of monitoring electrically the functioning of cam-operated switches, since failure is often caused by a spring fracture, or breaking of the arm, or loss of the roller follower.

Where possible, vane-operated magnetic proximity switches should be used. They are totally enclosed with sealed contacts and have no mechanical

Figure 7.16. Pivot bearing beam and prestressing cable anchorage

moving parts. These switches provide a much higher degree of reliability than can be achieved with cam- or lever-operated limit switches. They can be supplied suitable for immersion in up to 100 m of water. The maintenance of accurate working clearances is important when proximity switches are used.

7.5. Ropes

The usual practice is to provide two ropes for hoisting and attach them to the gate by means of a load-compensating arm. The arm is arranged so that in the event of failure of a rope, hoisting can continue with the other rope. Stainless steel ropes are available but usually in limited sizes and rope construction only. Spearman[2] records electrolytic corrosion of stainless steel ropes of spillway gates where the ropes were located upstream of the gate skin plate.

Ropes are subject to elongation due to the settlement of the wires in the strands and the strands in the rope. When geared limit switches are used, of the type that measure the distance of travel of a gate by the rotation of a hoisting drum, the limit switches have to be reset after the initial constructional extension of the rope has occurred. Alternatively the ropes can be supplied prestressed.

The modulus of elasticity of a rope is much less than that of the steel of the individual wires because of the helical winding of the wires making up a strand and the helix of the strands. It varies according to the construction of the rope. Table 7.2 gives some values.

Wire ropes should be pre-greased by the manufacturer and subsequently dressed at regular intervals. Generally in gate hoisting applications the factor of safety on the rope breaking strength is 8, although a load which results in a factor of safety of 5 produces what is considered a 'normally loaded' rope.

7.6. Chains

Two types of chain are used for hoisting of gates. In the roller chain shown in Figure 7.17(a) (known as a 'Galle' chain) the link pins rotate in the chain links and the link pins slide in the sprocket teeth during hoisting. In Figure 7.17(b) the link pin carries a bush. With this type of chain there is no sliding of the pin relative to the teeth of the sprocket; the movement occurs between the bush and the pin. The chain absorbs less power and the chain can be supplied with grease nipples in each pin so that all rotating faces can be lubricated. The chain in Figure 7.17(a) is usually lubricated by drip feed. Difficulties have been experienced with chains of type (a) due to high friction at the pin face which bears on the link. This has resulted in chains failing to articulate fully. This is less likely to happen with chains of type (b) and (c).

Table 7.2. Apparent modulus of elasticity of wire ropes[3]

Type of rope	E N/mm$^2 \times 10^3$
6 stranded ropes—fibre core, simple construction (e.g. 6×7)	62.0
6 stranded ropes—steel core, simple construction (e.g. 6×7)	68.7
6 stranded ropes—fibre core compound construction (e.g. 6×19, 6×36)	49.0
6 stranded ropes—steel core compound construction (e.g. 6×19, 6×36)	59.0
Multi-strand non-rotating construction (e.g. 17×7)	42.0

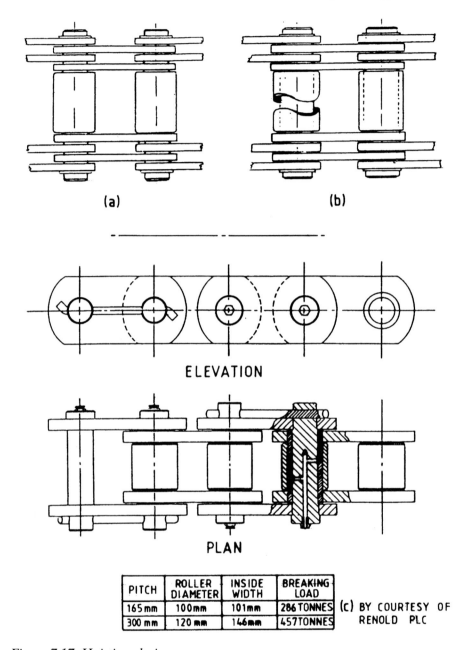

(a)   (b)

ELEVATION

PLAN

PITCH	ROLLER DIAMETER	INSIDE WIDTH	BREAKING LOAD
165mm	100mm	101mm	286 TONNES
300 mm	120 mm	146mm	457 TONNES

(c) BY COURTESY OF
RENOLD PLC

Figure 7.17. Hoisting chains

References

1. DIN 19704 (1976): *Hydraulic Steel Structures: Criteria for Design and Calculation*.
2. Spearman, P C (1967): Design and Development of Radial Spillway Gates in New Zealand, *New Zealand Engineering*, Feb.
3. Bridon Ropes (1992): *Steel Wire Ropes and Fittings*. Publication 1304.

8
Embedded parts

Sill beams, side-seal contact and roller faces on radial gates, gate roller and sliding paths for vertical-lift gates and tunnel lining sections of high-head gates have to be embedded in concrete. They have to be rigidly secured and accurately aligned. The practice is to provide cut-outs in the primary concrete and means of fixing alignment screws. The embedded parts are then accurately set up and secondary concrete is cast around them. To illustrate the sequence of erection an example of an embedded sill beam is shown in Figure 8.1.

The pads (1) for the adjusting studs are cast into the primary concrete. The adjusting studs (2) are then welded to the pads. This is followed by positioning the sill beam (3), aligning and levelling by adjusting the nuts on the studs. The final operation is to cast the secondary concrete. Adjusting studs should not be less than 15 mm in diameter. They should not be assumed to tie in the primary and secondary stage concrete. Separate reinforcement should be provided to carry out this function. Dovetailing the first-stage blockouts on the sides is advantageous.

The best practice is to machine the top flange of the sill beam and to line it with a stainless steel sill seal contact plate. The plate is either welded to the beam, or in some cases screwed to the beam so that it can be renewed. If this practice is adopted insulation against electrolytic corrosion between the carbon steel and the stainless steel is advisable.

Figure 8.2 shows an example of the embedded side-seal contact face for a radial gate and the roller path for the gate side-guide rollers. (1) is a rail for the gate transverse guide roller; (2) is the roller path; and (3) is the side-seal contact plate.

Figure 8.3 illustrates the gate slot of a high-head, vertical-lift roller gate. (1) is a rail for the gate transverse guide roller; (2) is the roller path; and (3) is the side seal contact plate.

In the last example, a rigid fixing and alignment is provided for the embedded parts of a lintel seal of a high-head slide gate (Figure 8.4).

It is highly desirable for all faces in contact with water to be of stainless

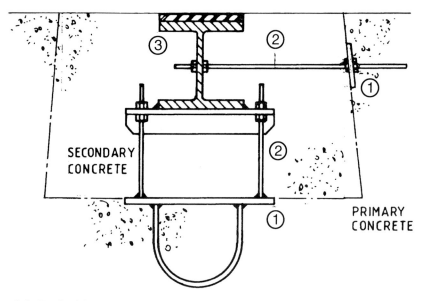

Figure 8.1. Embedded parts

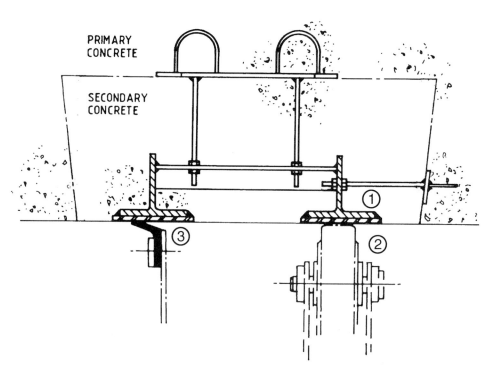

Figure 8.2. Side seal contact face and guide roller path for a radial gate

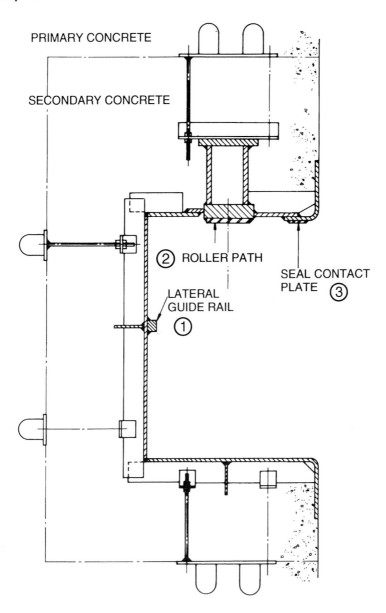

PRIMARY CONCRETE

SECONDARY CONCRETE

② ROLLER PATH

SEAL CONTACT PLATE ③

LATERAL GUIDE RAIL ①

Figure 8.3. Gate slot of a high head vertical roller gate

steel. The corrosion resistance of stainless steels depends on the alloying content of chromium and nickel. Therefore, the austenitic stainless steels with chromium content of 15% or greater and nickel content of 10% or greater have the best corrosion resistance of the three groups of stainless steel (see also Chapter 13). Unprotected low carbon steels should not be closer than 75 mm to a water face.

Steel linings in gate slots or tunnel inverts can be repainted when stoplogs

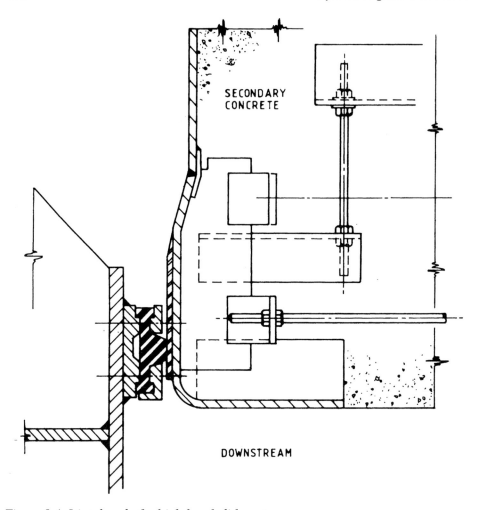

SECONDARY
CONCRETE

DOWNSTREAM

Figure 8.4. Lintel seal of a high head slide gate

or bulkhead gates have been positioned and the section has been dewatered. This does not apply to steel lining for stoplog slots which cannot be refurbished throughout the existence of the structure. In practice the application of stainless steel to faces in contact with water is often confined to seal contact and sliding faces.

The design criteria for thrust faces of embedded parts, sill beam and slide or roller paths are empirical. The distribution of load is assumed to be effected by the lower flange of the beams (Figure 8.5 (b)). The dimensions of the distribution cross-section are given in DIN 19704[1]. In a gate slot, the minimum distance from the outer edge of the concrete shall not, as a rule, be less than 150 mm.

The design of the beam is conventionally based on that of a beam on elastic foundation with the modulus of concrete having a value of $C = 200$ N/ mm^3.

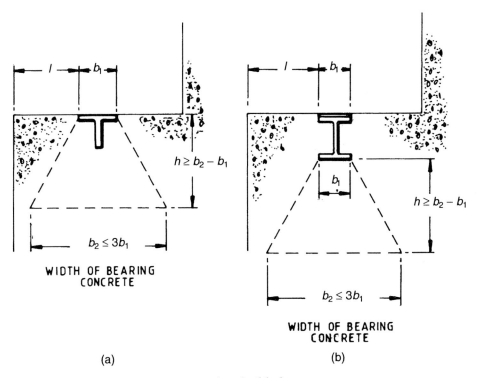

Figure 8.5. Concrete bearing area of embedded parts

The usual checks apply, such as the compressive stress of the concrete below the transmission area and the shear stress in the beam. Some designers advocate a corrosion allowance for all embedded parts of carbon steel. If the usual design practice for hydraulic equipment is followed (that is, derating the permissible working stresses) this can be accommodated within such an allowance.

At high-head gates the jet emitted under the gate at small openings can cavitate and become attached to the bottom of the conduit. It is therefore common practice to line the invert for 2 to 4 m downstream of the sill and to line upstream between half and two thirds of the downstream length.

In the situation where a hydraulic jump downstream of a gate is contained within a concrete tunnel, considerable erosion to the invert can occur due to recirculation of debris within the jump[2]. Under conditions of a hydraulic jump it may be necessary to extend the invert lining to protect the concrete.

References

1. DIN 19704 (1976): *Hydraulic Steel Structures; Criteria for Design and Calculation.*
2. Lewin, J; Whiting, J R (1986): Gates and Valves in Reservoir Low Level Outlets; Learning from Experience, *BNCOLD/IWES Conference on Reservoirs*, Edinburgh, Sept, p 77.

9
Hydraulic considerations pertaining to gates

The first section in this chapter deals with the basic data required to determine the stage discharge characteristics of gates. The discharge coefficients for radial gates used in the equations are not directly comparable because the definition of energy head varies. This ranges from the head upstream of the gate to the head to the middle of the gate opening and in drowned-discharge both the true energy head, the difference between the upstream and the downstream water levels, and the downstream head enter the equation.

Little information has been published on the stage discharge relationship of top-hinged flap gates. Available data have therefore been included in this chapter.

The next section deals with hydraulic downpull forces on vertical-lift gates. This hydrodynamic effect is usually ignored when designing gates in open channels because under these conditions it is low and is absorbed by the margin of hoisting force provided in gate installations. For high-head gates it becomes important. Because of the number of variables involved in determining hydraulic downpull, calculations must be considered approximate. All research in this field has been carried out on models representing vertical-lift gates, although hydraulic downpull forces also act on radial gates.

Attention is drawn next to instability in a reach of a watercourse which can be caused by the operation of a gate when there is limited ponded up water. Problems can also arise when there is a change from three- to two-dimensional flow. This is the condition of flood flow from a reservoir into a sluiceway. This type of problem can be resolved by a physical model study.

The occurrence of reflux at a multi-gate installation and observed flow oscillations are described, as well as the hysteresis effect of gate discharge during raising and lowering.

The section on hydraulics considerations pertaining to gates in conduits deals initially with vorticity at intakes. While this is a matter of design of the civil engineering structure, the introduction of air into a conduit can cause severe pressure fluctuations at control gates. Awareness that free vortices can occur should result in reconsideration of the design of an intake.

Cavitation and erosion are factors whenever flow velocities of the order of 13–15 m/s are reached or exceeded. Cavitation can affect gate slots and the invert of tunnels. Information relevant to gates is included in this section and is complementary to that on cavitation in valves (Chapter 3).

Other hydraulic considerations complete the chapter. These are pressure coefficients for gate slots, confluence of jets created by gates in parallel conduits, proximity of two gates in parallel and air demand.

9.1. Flow under and over gates

9.1.1. Flow under gates

The coefficient of discharge of a radial gate installed in an open channel watercourse to control water level and flow rate varies as the gate geometry, the opening and the upstream and downstream water levels. For submerged discharge it ranges from 0.3 to 0.6 and for free discharge from 0.5 to 0.7. Rouse[1] gives a graph based on Metzler[2] for one value of the gate radius to height of pivot above the sill. This is reproduced by Lewin[3]. Chow[4] gives discharge coefficients for radial gates where the radius of the gate is the denominator in the non-dimensional functions, whereas Metzler[2] uses the height of the pivot above the gate sill.

For a flat, vertical, sluice gate Franke and Valentin[5] developed a discharge formula for free flow by measuring the pressure at the floor directly below the gate lip and relating this value to the geometry of the jet. The pressure can be determined analytically and Franke and Valentin developed an expression for the free discharge case. Young and Fellerman[6] extended this for the general case of submerged flow. In many cases a direct solution can be obtained from the expression for the general case; in others, a solution has to be obtained by trial and error.

Should the floor level drop appreciably downstream of the sluice gate, the equation derived by Young and Fellerman cannot be applied and direct pressure measurements are then required. Another limitation arises when the jet efflux at the gate opening attains a sub-critical value.

The general equation for discharge through an underflow gate can be expressed as:

$$Q = C_d \times Go \times W \sqrt{(2gH)} \qquad\qquad 9.1$$

where Q = discharge
C_d = coefficient of discharge
Go = gate opening = b
W = gate width
g = acceleration due to gravity
H = upstream water head

The variables affecting the discharge characteristics of a radial gate are shown in Figure 9.1.

An example of the coefficient of discharge map for free and submerged flow under a radial gate based on Metzler[2] is shown in Figure 9.2.

Buyalski[7] used similar maps to derive discharge algorithms which can be programmed into a computer for automatic control of gates or for calculating discharge based on measurement of water levels and gate attitude, converted to gate opening. The algorithms derived by Buyalski were based on experiments carried out with gates having a lip seal of hard rubber with rectangular section. The seals were mounted upstream of the gate skin plate, whereas the preferred practice is to locate the seal downstream of the skin plate. An upstream seal causes a flow disturbance and is not in accordance with the rule suggested by Lewin[8] and Vrijer[9] to the effect that flow separation should be arranged at the extreme downstream edge of a gate to achieve flow conditions that are as steady as possible. Buyalski[7] states that the experimental data show that different gate lip designs (even a minor modification) can result in a −7% to +12% difference in the coefficient of discharge C_d. The different seal configurations investigated were the rectangular hard rubber section, a seal of musical note shape and no seal, that is metal edge contact. A seal of musical note shape should not be used as a lip seal as it can lead to gate vibration.

The US Corps of Engineers Hydraulic Design Criteria[10] include graphs for free discharge for ratios $a/r = 0.1, 0.5$ and 0.9, where a is the height of the gate pivot above the sluiceway floor and r is the radius of curvature of the skin plate. The graphs are based on Toch[11], Metzler[2] and Gentilini[12]. However the geometry of many gates is outside the range of these ratios. The charts are useful because they incorporate an ancillary graph to give adjustment factors when the gate sill is raised above the floor of the channel.

The US Corps of Engineers'[13] chart for the coefficient of submerged

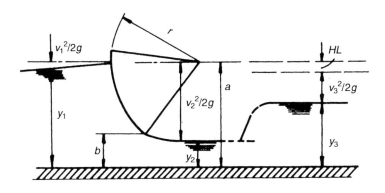

Figure 9.1. Variables affecting the discharge characteristics under a radial gate

discharge is independent of the *a/r* ratio and is plotted for the ratio of a raised sill with downstream submergence over the gate opening. It appears that the height of the sill above the approach bed is not an important factor in submerged flow controlled by gates. One of the graphs (sheet 320–8) is reproduced in Figure 9.3.

For radial gates on spillway crests, the discharge through a partially open gate can be computed using the same basic orifice equation:

$$Q = CA\sqrt{(2gH)}$$ 9.2

where C =   coefficient of discharge
         A =   area of opening
         H =   head to the centre of opening

The coefficient is primarily dependent upon the characteristics of the flow lines approaching and leaving the orifice. In turn, these flowlines are dependent upon the shape of the crest, the radius of the gate and the location of the gate pivot.

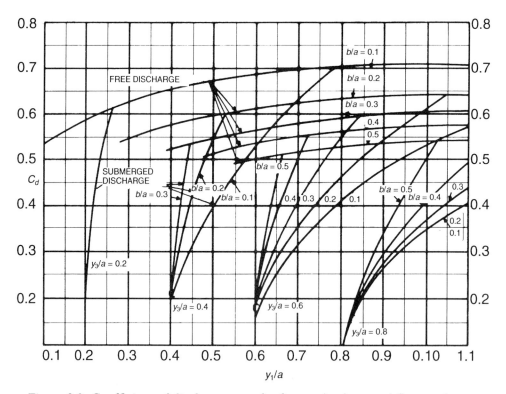

Figure 9.2. Coefficient of discharge map for free and submerged flow under a radial gate for r/a = 1.15

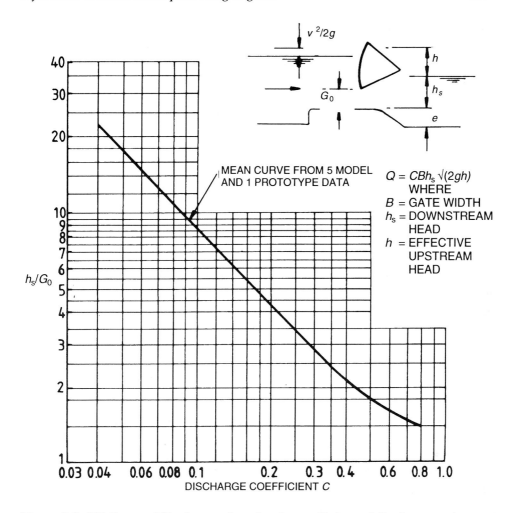

Figure 9.3. *US Corps of Engineers chart for the coefficient of discharge under submerged flow conditions*

The Hydraulic Design Criteria[14] plot average discharge coefficients from model and prototype data for several crest shapes and gate designs for non-submerged flow. On the chart, the discharge coefficient is plotted as a function of the angle (β) formed by the tangent of the gate lip and the tangent to the crest curve at the nearest point of the crest curve. This angle is a function of the major geometric factors affecting the flow lines of the discharge. Figure 9.4 gives suggested design values for discharge coefficients of 0.67 to 0.73 for β from 50° to 110°.

The seat or sill of radial gates on spillway crests is usually located downstream of the crest axis. Provision is made for placing of stoplogs upstream of the gates. Placing the gate seat and the seat for the stoplogs close to the crest axis reduces the overall height of the gates and the stoplogs in

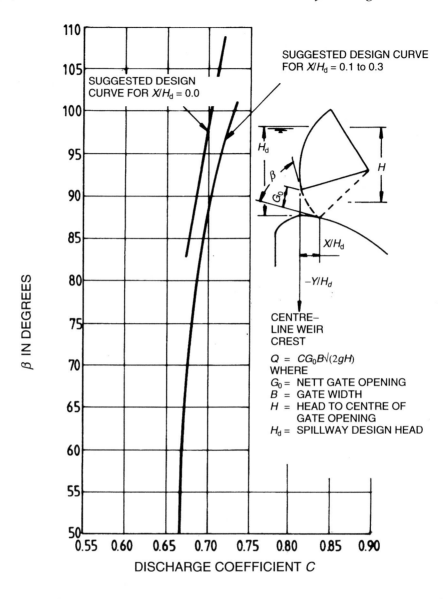

Figure 9.4. Coefficient of discharge for radial gates on spillway crests for gate lip angle from 50° to 110°

relation to the reservoir retention level. A practice favoured by gate designers is therefore to make the gap between stoplogs and gate just sufficient, so that work can be carried out within the space with the stoplogs located upstream of the crest axis and the gates downstream. Limited test results suggest that within the normal practical dimensions of location of the gate sill there is no effect on the discharge coefficient, but the crest pressure will be affected[15]. Slight negative pressures occur on the spillway crest for a

G_0/H_d ratio of 0.4 or with the gate seat located on the crest axis. Crest pressures derived from the US Corps of Engineers'[15] charts are positive for all other G_0/H_d ratios and gate seats downstream of the crest axis.

The discharge coefficients in US Corps of Engineers[10,13,14] are based principally on tests with several bays in operation and it is suggested that discharge coefficients for a single bay would be lower because of side contraction. Limited experimental data[16] indicate that provided the gate piers project at least half a bay width upstream of the gate sill and the approach channel is sensibly straight, each sluiceway operates as if it were independent of the adjoining bays. The requirement for projection of the piers is usually met because of the practice of locating a bridge upstream of the gates for access purposes and for mounting a gantry crane for handling of stoplogs. To compute the discharge through each bay, the pier flow contraction coefficient[17] must also be considered.

The equation of discharge through a vertical-lift gate is the same as Equation 9.1 for a radial gate. Using the same notation as Equation 9.1, the variables affecting the discharge characteristics are shown in Figure 9.5 based on Rouse[1].

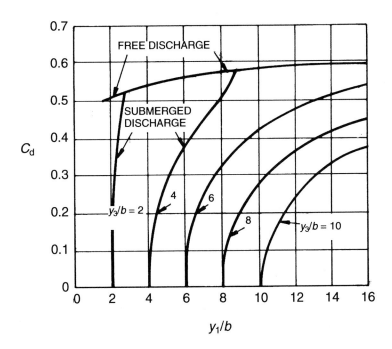

Figure 9.5. Coefficient of discharge map for free and submerged flow under a vertical lift gate

9.1.2. Flow over gates

Radial gates have been designed to be overtopped. The top of the gate then acts as a sharp-crested weir. If the nappe impacts on transverse structure stiffener beams downstream of the skin plate, it can result in gate vibration. When a radial gate with an adjustable overflow section discharges over the gate, the flow conditions are those of a broad-crested weir and the shape of the crest is designed to avoid negative pressure and flow separation. This also applies to hook gates. Bottom-hinged flap gates can be subject to flow conditions of a broad-crested weir changing with increasing elevation of the gate to those of a sharp crested-weir.

Drum, sector and bear-trap gates are all subject to varying discharge coefficients throughout their raising and lowering motions.

The major load acting on an overflow gate or bottom-hinged section of a gate is the hydrodynamic water pressure. The pressure is dependent on the velocity distribution of the flowing water, the shape of the flow boundaries, the separation of the water stream from, or its adhesion to, the gate surface. The hydrodynamic pressure is of a pulsing character due to random velocity pulsations[18,19], the instability of the flow separation point from the gate and water level oscillations induced by wave motion or a change in flow conditions. The distribution of mean and fluctuating pressures for different operating conditions is usually determined by a model study[20].

An analytical method of determining the mean value of local hydrodynamic pressure and a mathematical model of pressure pulsation for a non-submerged bottom-hinged gate with a circular curvature is given by Rogala and Winter[21], who derived an equation to determine the hydrodynamic pressure at an arbitrarily chosen point on an overflow hinged gate which is not submerged and fully aerated. The pressure at a point depends on the geometric parameters of the gate, its position, and the flow discharge.

The equation is:

$$\frac{\bar{p}}{\rho g R} = \frac{H_0 - y}{R} - 0.85 \times F \times \exp{(W)} \qquad 9.3$$

where $\bar{p}$ = mean pressure at the point examined on the gate surface
      ρ = water density
      g = acceleration due to gravity
      H_0 = heights of the energy head above the gate hinge
      y = vertical distance above gate hinge
      R = radius of gate curvature
      F = v^2/gR where v = mean horizontal velocity of flow over gate edge)

W = an exponent dependent on the gate inclination angle α and on the angle co-ordinate Φ of the point examined on the gate, as well as $H_0 - y$ and h_0, calculated from Equations 9.4 and 9.5.

$$W = 1.22\,Y^{0.7} \text{ for } \alpha < (\Psi/2) \text{ and } \Phi > (\Psi/2)+\alpha \qquad 9.4$$

$$W = -7.1\,F^{0.5}\,Y^{0.4} \text{ in the remaining cases} \qquad 9.5$$

$$Y = \frac{(H_0-y)/R}{(h_0/R)-1} \qquad 9.6$$

9.1.3. Stage–discharge relationship of a top-hinged flap gate

A theoretical treatment of the stage–discharge relationship of a rectangular flap gate was established by Pethick and Harrison[22]. This was derived from two different theoretical concepts which yielded sensibly similar results.

(a) Free flow

For free flow the flap gate relationship was expressed as a three dimensionless parameter plot, Figure 9.7.

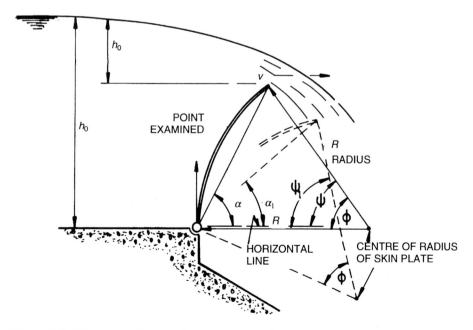

Figure 9.6. Diagram of water flow over hinged gate

where q = discharge per unit width of flap gate
       ρ = density of fluid
       h = upstream head
       L = depth of flap
       βL = distance from hinge to the centre of gravity of the flap
       m = mass per unit width of flap gate
       g = gravity constant

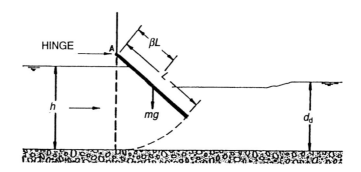

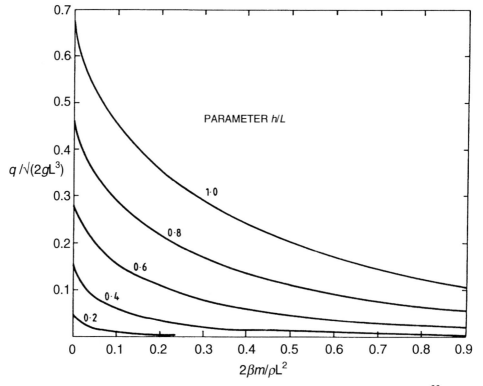

Figure 9.7. Rectangular flap gate—free flow (after Pethick and Harrison[22])

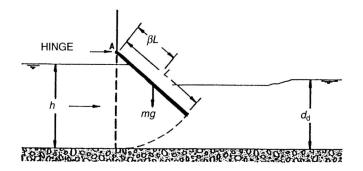

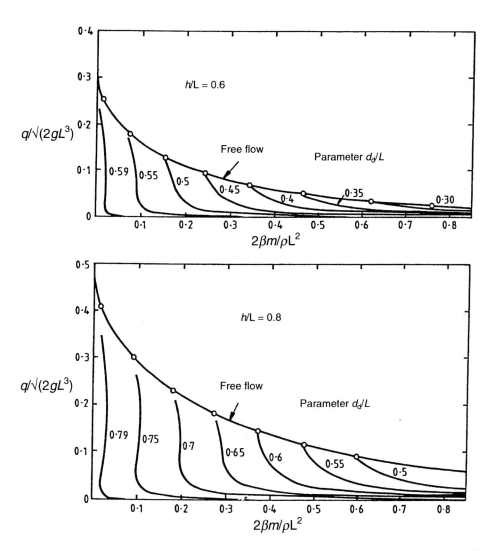

Figure 9.8. Rectangular flap gate—drowned flow (after Pethick and Harrison[22])

For a given gate, the mass parameter $2\beta m/\rho L^2$ is constant and thus the stage–discharge relationship is read from the ordinate in Figure 9.7 and h/L values on the appropriate vertical line. For design purposes the mass parameter can be determined if the values q, h and L are known. In practice, h is likely to be variable and design will have to be carried out on a trial and error basis.

Pethick and Harrison[22] point out that the Froude number upstream of the gate $q/\sqrt{(gh^3)}$ is unity at the point where the h/L curve intersects the vertical axis.

(b) Drowned flow

The drowned flow leads to a four parameter relationship shown in Figure 9.8. It is plotted for h/L values of 0.6 and 0.8.

The theoretical analysis does not give a solution over a small range close to the free flow limit. Visual observation of model tests suggest that there is a discontinuity due to unstable flow in this range.

The figures show that throughout the drowned flow regime for a constant upstream water level, the discharge reduces very rapidly as tailwater increases; $h - d_d$ is the effective head across the gate and, as expected, q is proportional to $\sqrt{(h - d_d)}$.

(c) Empirical stage–discharge relationship

Tests carried out on circular flap valves at the State University of Iowa, USA, to establish loss of head through flap valves derived the following empirical formula:

$$L = \frac{4V^2}{g} \times \exp\left(\frac{-1.15V}{\sqrt{d}}\right) \qquad\qquad 9.7$$

where L = loss of head
V = velocity of flow through the gate in feet per second
d = diameter of the outlet in feet

It is assumed that this formula applies only to free discharge although this was not stated.

9.2. Hydraulic downpull forces

Hydraulic downpull forces under a gate are due to the reduction in pressure caused by the discharge under the gate. For gates in open channels this is the main factor affecting downpull and is a function of the gate geometry. Approach flow effects can modify the downpull effect slightly.

In tunnel gates and particularly in high-head gates, the downpull is significantly affected not only by the geometry of the gate bottom, but also

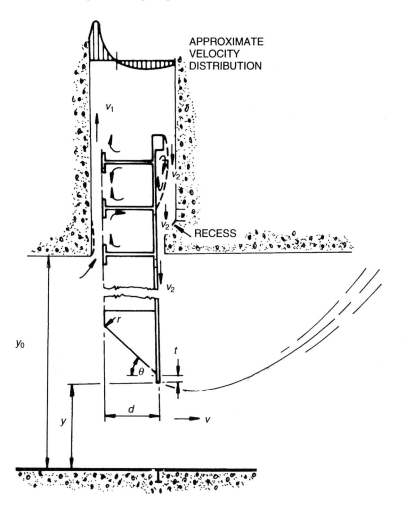

Figure 9.9. Diagram of tunnel gate under submerged flow conditions (after Naudascher and Locher 1986[23])

by the rate of flow passing over the top of the gate through the gate well, which can exert a major effect on the magnitude of the downpull.

There are two states of flow at a high-level gate:

- Free flow, in which the space downstream of the gate is filled with air.
- Submerged flow, in which that space is submerged and pressurised (Figure 9.9, after Naudascher and Locher[23]).

The flow conditions for a typical arrangement of a gate partially withdrawn into a well are also illustrated in Figure 9.9.

The primary part of the downpull stems from the difference of the integrated distributions of piezometric head on the top of the gate and the bottom surface and may be expressed as:

$$F = (\alpha_T - \alpha_B) \, Bd\rho \, (V^2/2) \qquad\qquad 9.8$$

where F = downpull force
        α_T = downpull coefficient at the top of the gate
        α_B = downpull coefficient at the bottom of the gate
        B = width of gate
        d = depth of gate
        ρ = density of water
        V = velocity in the contracted section of the jet

The geometric parameters which affect the downpull at the bottom of the gate are:

- The ratio of gate opening to the tunnel height which can be expressed as y/y_0 the percentage opening of the gate.
- The angle of the gate bottom θ
- The ratio of the radius r to the depth of the gate r/d
- The ratio of the projection of the gate lip t to the depth of the gate t/d
- The relative conduit height y_0/d

It is usual to make the angle $\theta = 45°$ for stiffness of construction and because it is favourable from considerations of downpull force, although one model study[25] showed that a reduction of 5% in downpull can be achieved by increasing θ to 50°.

A radius r at the transition is important and prevents flow separation at the lower end of the vertical face. Design practice is to make t a minimum. Further downpull over and above F is due to the pressure difference acting on the horizontal projection of the top seal of the gate or the projection of the extended skin plate. When the gate lip approaches the tunnel ceiling, the downpull may become negative, i.e. transform into an upthrust. Under those conditions it could inhibit safe gate closure.

Figures 9.10 and 9.11 show downpull coefficients for tunnel-type gates versus gate openings. The graphs incorporate the results of two different investigations. When using these coefficients an estimate has to be made of the contracted jet issuing from beneath the gate. The discharge coefficient of the jet varies with gate opening and gate geometry and is superimposed on Figure 9.10.

Figure 9.12[25] illustrates the dependence of the bottom downpull coefficient on gate geometry and Figure 9.13[25] on the relative conduit height. Naudascher et al[23] conclude that gate slots do not affect hydraulic downpull forces.

At intake gates the approach flow conditions can cause strong variations of downpull and discharge coefficients[25]. Piers at the intake and trash rack grids, a short distance upstream of the gate, can cause flow separation or

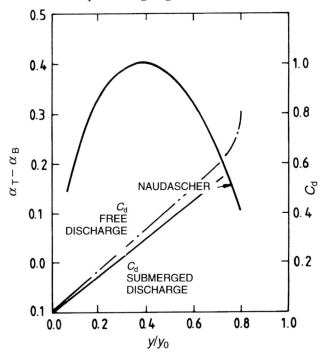

Figure 9.10. Resultant downpull and discharge coefficients versus gate opening

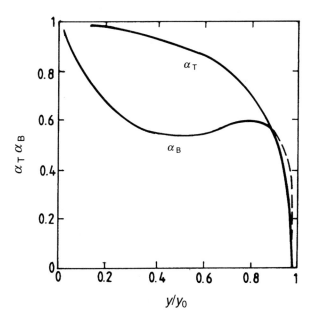

Figure 9.11. Top and bottom downpull coefficients versus gate opening (after Weaver and Martin[24])

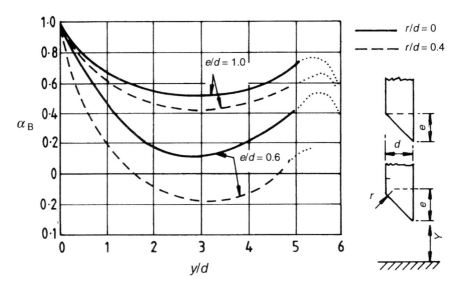

Figure 9.12. Dependence of bottom downpull coefficient on gate geometry (for a ratio of conduit height to the depth of the gate. $y_0/d = 6$ (after Thang and Naudascher[25])

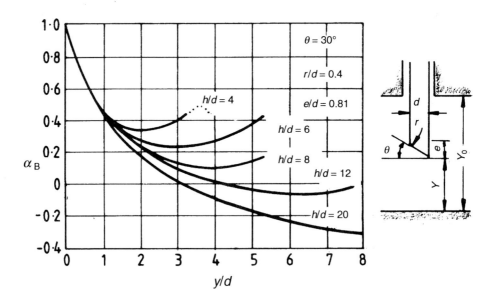

Figure 9.13. Dependence of bottom downpull coefficient on relative conduit height y/d (after Thang and Naudascher[25])

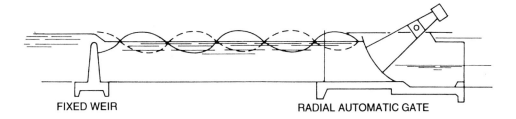

FIXED WEIR   RADIAL AUTOMATIC GATE

Figure 9.14. Wave action due to limited ponded-up water

alter the turbulence characteristics of the near gate flow regime compared with free stream turbulence.

Since several variables are involved in hydraulic downpull forces, the values of the downpull coefficients shown in the graphs should be considered indicative and not actual design values. It may sometimes be expedient and cost effective to oversize the servo-motor by assuming sub-atmospheric pressure at the gate bottom.

9.3. Limited ponded-up water

An opening of a gate controlling a limited reach of a river will send a wave upstream. When this wave is reflected it can, in turn, cause a disturbance of the gate. The motion can amplify and cause serious instability (Figure 9.14). An instance of the type of instability of a radial automatic gate is cited by Lewin[8].

A change in the discharge under or over a gate will similarly cause a wave to travel upstream and if it is reflected it will register a false increase in water level. This can actuate the control system and therefore initiate a further opening of the gate or gates.

9.4. Three-dimensional flow entry into sluiceways

Where the flood flow from a reservoir into sluiceways is not trained by an approach channel, cross-flow will occur (Figure 9.15). This can result in eddy shedding at the piers and turbulent conditions at the gate face.

Kolkman[26] quotes an example of self-exciting wave oscillations in the upstream basin of a sluice experienced during a model study carried out in the Delft Laboratory. When, for instance, only six openings out of ten discharged, while the others (and especially the outer ones) were closed by gates, a transverse wave oscillation occurred, resulting in a wave amplitude in the prototype of 2 m near the closed gates. The transverse flow component related to the water level oscillations most probably interacted at the point where the main flow separated from the side walls.

Figure 9.15. Cross-flow due to change from three-dimensional to two-dimensional flow

9.5. Reflux downstream at a pier and flow oscillation

In multi-gate operation, the drowned-discharge from one gate, when the adjoining gate is closed, can reflux into the quiescent stilling basin and cause a periodic disturbance. This can act on submerged structural members downstream of the gate and can result in gate oscillations. In an installation of two radial automatic gates, the pier between the gates extended 15.75 m downstream of the sill and had to be extended by an additional 8 m to prevent gate oscillation under the conditions previously described.

Flow oscillation has occurred when the water surface differential between pool and tailwater level is relatively small and the flow is controlled primarily by the tailwater[27]. It caused bouncing of the radial gates due to surges of flow which moved back into the gate bay and struck the bottom girder of the gates. The fluid flow was strong enough to lift the gates causing the bouncing phenomenon. Extension of the piers would have reduced the load on the gates. Flow oscillation has also been noted in other model studies[28].

9.6. Hysteresis effect of gate discharge during hoisting/lowering

When flow under a gate becomes detached from the gate lip it accelerates and drops away from the gate. To resume control of the flow by the gate it has to be lowered further. Contact of the gate lip with the face of the water will initially have no effect. A slight further immersion will cause an afflux at the gate and an increase in upstream water level higher than the level prior to the upstream flow becoming detached. Where this sequence of events can affect level measurement for a control system, it can result in unstable operation.

9.7. Hydraulic considerations pertaining to gates in conduit

9.7.1. Vorticity at intakes

When air is introduced into a conduit, severe pressure fluctuations can occur at a control gate due to the build up of stagnated air under high pressure at the conduit crown upstream of the gate. This effect is discussed in Chapter 10 under Two-Phase Flow. The formation of free vortices at intakes must therefore be avoided.

Two factors principally determine the formation of vorticity at an intake, submergence and circulation of the approach flow. Circulation is the primary parameter to influence submergence[29,30]. Gulliver *et al.*[31] have suggested that, as a first approximation prior to a model study, two design parameters should be used, the dimensionless submergence $S/D = 0.7$ to 4 (Figure 9.16) and Froude number $F_r = 0$ to 0.5. Intakes at existing stations and model studies within these limits experienced no vortex problems. It is presumed that the data to support this were obtained under reasonably straight approach flow conditions.

Gordon[32] gives the minimum submergence of the top of the gate opening as determined by the formula:

$$S = C \times v \times D^{1/2} \qquad\qquad 9.9$$

where C is a constant, suggested to be 0.3 for symmetrical approach flow.

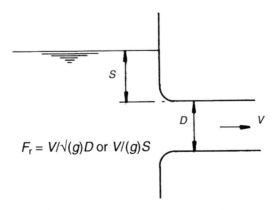

$$F_r = V/\sqrt{(g)D} \text{ or } V/(g)S$$

Figure 9.16. Vortex formation at a submerged intake

The units of C are $\text{sm}^{-1/2}$. The formula has been derived from observations on 29 prototype intakes. Intake screens will, by streaming the flow, permit lower submergence of inlets before the onset of formation of free vortices[33,34].

Anti-vortex devices have been used, such as long approach channel walls over the intake, or the hooded inlet developed by Song[35] and Blaisdell and Donelly[36].

9.7.2. Cavitation and erosion

The causes of cavitation and its effect on engineering structures have been well documented. The immense damage to the tunnels at Tarbela[17] (Chapter 14) in 1974 has been attributed to cavitation. With high velocities and the potential for sheared flow adjacent to conduit boundaries, regions of low pressure can be set up with pressures close to that of incipient cavitation. Small surface irregularities can be sufficient to drop the pressure to a level to initiate cavitation.

Considerable cavitation damage was reported by Wagner[37] due to high velocity flow up to 41 m/s. This was due to poor alignment of the liner joints, projecting joint welds and minor ridges and depressions in the paint coating. Offsets as little as 0.8 mm into the flow produced marked damage and the degree of damage increased with larger offsets. Depressed surface offsets of 6 mm produced paint removal and minor pitting. Laboratory studies were conducted by Ball[38] to establish the velocity–pressure relationship for incipient cavitation at offsets with rounded corners and sloping surfaces that protrude into the flow. These may be used as guidelines for establishing tolerances for surface irregularities of linings downstream of gates. The greater resistance of stainless and high nickel steels to cavitation damage is documented by Wagner[37] and others, although cavitation damage to stainless steel occurred downstream of the regulating gates at the Dartmouth Dam low-level outlet[39].

Cavitation conditions can arise at pier nosing where piers divide several gate passages, especially under asymmetric gate operating conditions. Anastassi[40] gives an equation for an elliptical shape of the pier nosing to reduce pressure fluctuations for symmetrical flow and a modified equation for a small elliptical shape for asymmetric flow conditions. The paper also notes that the taper of the pier must be gradual to prevent flow separation.

Serious cavitation can be caused by high-pressure flow through small gaps at seals and at gates which are just closing or opening. Tests have shown that gaps of less than 0.1 mm are safe for short periods, whereas gaps of more than 2 mm can cause serious erosion, apart from the possibility of inducing gate vibration. Since cavitation damage is time dependent, high-head gates should not be kept in operation at small openings. Minimum openings should not be less than 100 mm. This will also minimise erosion downstream of the sill (see also 'Unstable Flow through Small Openings').

9.7.3. Gate slots

Vertical-lift gates of the roller or slide type require recessed slots in abutments or piers for the movement of the gate guide rollers or slides. The flow of water across the slots causes flow separation at the upstream edge of the slot and reattachment on the downstream side. Eddies are set up within the slots and vortices are formed. Under conditions of high velocity flow cavitation can occur within gate slots.

Flow conditions due to gate slots are influenced by the upstream and downstream edge shape and the cavity depth to length ratio (Figure 9.17). Radiusing the upstream edge should be avoided. A radius on the downstream edge reduces the energy dissipation. A single, stable vortex forms in cavities of d/w ratio close to unity. This results in low losses. Between d/w ratios of 0.2 and 0.8 circulation is unstable with periodic disturbances influencing the main flow. Loss coefficients for sharp-edged gate slots have a minimum of about 0.01 with d/w ratios of 0.5 which rises to 0.03 for d/w of 2.5. The loss coefficient is defined as the ratio of head loss $V^2/2g$ where V is the mean velocity.

The flow past the gate slot results in a reduction in pressure on the conduit wall immediately downstream from the slot. Cavitation can occur within the slot or downstream from the slot when high velocity flow occurs and there is insufficient pressure in the region of the slot.

In Chapter 3 cavitation in valves was discussed and varying intensities of cavitation were differentiated. These are:

* Incipient cavitation is the onset of the phenomenon and usually occurs intermittently over a restricted area. Noise is slight and there is no damage except at isolated local conditions, such as a step.
* The next stage is critical cavitation where noise and vibration are

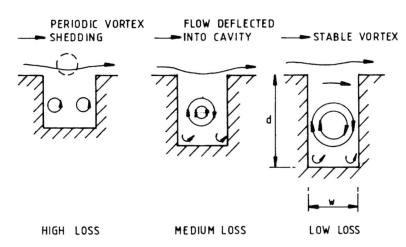

Figure 9.17. Flow in gate slots

acceptable and damage will occur only after long periods of operation. This is usually adopted as a design criterion in gate and valve installations.

• Further stage is incipient damage when pitting occurs after short periods of operation and is accompanied by a high noise level.

• Choking cavitation occurs when the outlet pressure is lowered to vapour pressure. At this stage the flow is unaffected by the downstream pressure, and flow and pressure loss relationships no longer apply. Close to choking, noise, vibration and damage due to pitting are at a maximum.

For the stage beyond choking cavitation, super cavitation, specialist literature should be consulted.

Cavitation of gate slots was investigated by Ball[41] and Galperin[42]. May[43] reviews cavitation in hydraulic structures and deals extensively with cavitation due to gate slots.

The cavitation parameter σ_s of a slot is given by:

$$\sigma_s = \frac{h_i - h_v}{v^2 / 2g} \qquad\qquad 9.10$$

where h_i = head

h_v = vapour head

Cavitation can be initiated by decreasing h_i or increasing v. Therefore the lower the cavitation parameter, the greater the intensity of cavitation.

If incipient cavitation is the design criterion and if the incipient cavitation parameter for a gate slot σ_{si} is known, the flow velocity at the gate slot can be calculated. A greater velocity will move the condition into the region of incipient damage cavitation.

In gate slots there are a number of geometric factors which affect the incipient cavitation parameter. These can be combined to form an overall value of σ_{si} in Equation 9.11. Figure 9.18 (after May[43]) shows the factors C_1, C_2, C_3 and K_{is} and their dependence on the geometry of a gate slot.

The incipient cavitation parameter for a gate slot is

$$\sigma_{si} = C_1 \, C_2 \, C_3 \, K_{is} \qquad\qquad 9.11$$

where K_{is} is the value of incipient cavitation at the upstream or the downstream edge of a gate slot.

In the graph of C_3 (Figure 9.18), σ is the thickness of the boundary layer which can be calculated from the boundary layer equation for smooth turbulent flow[43]. Since gate slots on high-head gates are long (w) in relation to the boundary layer thickness (σ), using a value of 1.4 for C_3 will result in safe designs.

The results are applicable to a fully open gate and when the flow is

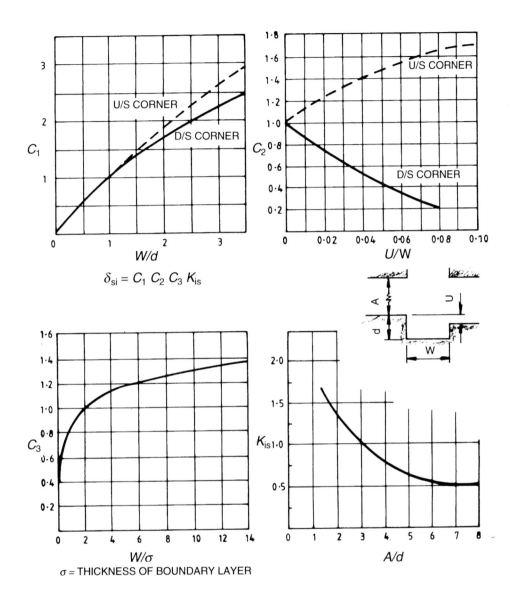

Figure 9.18. *Factors for incipient cavitation parameters of gate slots (after May[43])*

approximately two-dimensional. The latter condition may not apply to an intake gate.

Galperin[42] also gives data for vertical-lift gates which are partially open. Typical values of σ_{si} for gates discharging under submerged conditions can vary between 1.0 at a gate opening of 35% and 2.5 at 90% open. For free discharge the range is 0.3–1.0.

Another important variable is the conduit geometry downstream of the slot. The low pressure conditions on the downstream edge of the gate slot can be improved to some degree by offsetting the downstream edge of the slot and returning gradually to the original conduit wall alignment. Figure 9.19 shows pressure coefficients for gate slots with and without downstream offsets and with downstream rounded corners. The coefficients were computed using the equation

$$H_d = CH_v \qquad\qquad 9.12$$

where H_d = pressure difference from reference pressure
       C = pressure coefficient
       H_v = conduit velocity head at reference pressure

Figure 9.17 illustrated flow patterns in the plane of the slot. In addition forced vorticity occurs in the vertical direction resulting in a complex three-dimensional flow when the gate is in the open or partially open position. The pressure coefficients given in Figure 9.19 therefore illustrate only the improvement which can be achieved by providing a downstream offset. There are likely to be significant resultant hydraulic forces on the gate rail, tending to lift it from its mounting, resulting from high stagnation pressures which can be developed near the downstream edge of the slot where the gate rail is fastened.

Ball[41] showed that deflectors at the upstream edges of slots produce an ejector action which lowers the pressures at the slot far below the reference pressure and will induce cavitation. A very large deflector which causes a heavy contraction can be used successfully and is the basis of the design of jet-flow gates (see 'Control Gates and Guard Gates', Chapter 2). Some of the conclusions of the paper by Ball[41] can be used as a guide for the design of gate slots. Offset corners of slots and a variable rate of convergence are most desirable from hydraulic considerations. Arcs used in this design should have radii in the range of about 100 to 250 times the offset of the downstream corner. Ellipses can also be used with excellent results.   The upstream corners of the gate slots should not be rounded or notched, both are detrimental to the pressure distribution.

The widening of slots permits more expansion of the jet into the slot, tending to increase the contraction at the downstream corner. However, pressure conditions are acceptable for a wide range of slot width to depth ratios in designs using offset corners with converging walls. This is particularly true for the 24:1 convergence and the long radius curved convergence.

Sharp downstream corners of gate slots should always be offset away from the flow. The offset of the downstream corner of a gate slot should be small and related to the slot width. Within reasonable limits, this offset is not critical. Abrupt offsets into the flow and irregularities in flow surfaces are particularly troublesome. Offsets of less than 3 mm will cause damage. It is

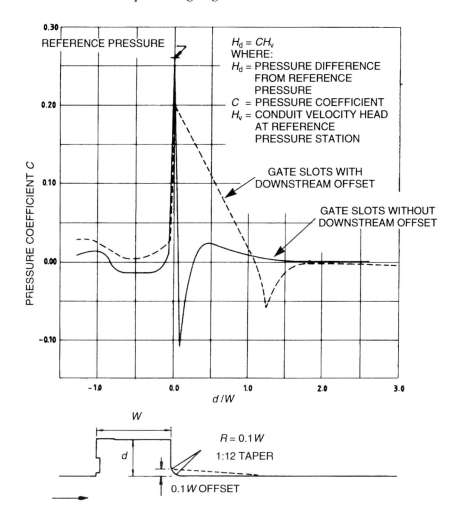

Figure 9.19. Pressure coefficients for gate slots with and without downstream offsets

extremely important to provide smooth continuous surfaces downstream from gates operating under high-heads.

9.7.4. Gate conduits

The investigation of the bottom outlet of the San Roque Dam in the Philippines[40] demonstrated severe turbulent flow separation upstream of the control gate installation of the type illustrated in Figure 9.20. This was of a periodic nature causing peak pressure surges. The geometry of the approach section of the tunnel and the transition to the conduits containing the gates affected the pressures in the gate chamber.

This paper also drew attention to the desirability that the cross-sectional

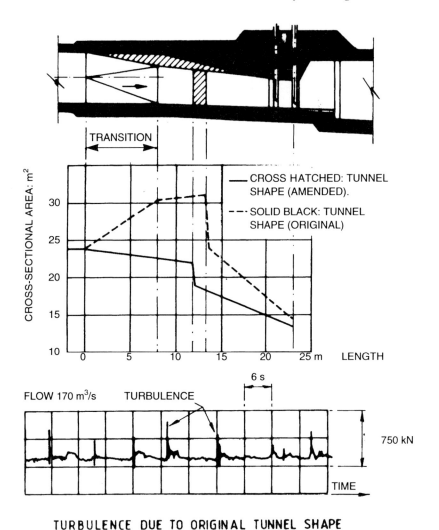

Figure 9.20. Bottom outlet of the San Roque Dam — flow separation and turbulence within the chamber upstream of the gate fluidways

area of the conduit at the point of gate discharge should be less than that of the approach tunnel to avoid sub-atmospheric pressures which could limit the opening of the control gate.

9.7.5. Confluence of jets created by gates in parallel conduits

Where two or more gates are to be installed in parallel then it is necessary to consider the effects brought about by the conveyance of the jets downstream and of any possible combination of the jet downstream and of asymmetrical flow. Problems can result from flow separation, unstable flow, excessive bulking, oblique flow and cross waves.

Koch[44], in the model study of the bottom outlet of the Randenigala Project in Sri Lanka, found that a downstream length of 8 m was insufficient for the dividing wall. With velocities up to 43.2 m/s flow was separating from the curved face of the dividing wall. In order to guard against low pressures which were likely to result in cavitation, it was necessary to extend and taper the wall by 35 m and incorporate facilities for air entrainment.

The bottom outlet of the Mrica Hydroelectric Project[45] has twin conduits, each housing a control and emergency closure gate of the slide type. The dividing wall extends 8.8 m downstream of the sill of the control gate with no physical flow separation beyond the wall. The maximum jet velocity was 33.55 m/s.

9.7.6. Trajectory of jets due to floor offsets

The model study of the drawdown culvert control structure for Mrica[45] showed that the deflectors of the aeration slots in the invert downstream of the gates caused the trajectory of the jet leaving the step to be thrown up to strike the tunnel roof. Omitting the deflectors and modifying the step led to an acceptable trajectory but with a small decrease in the volume of entrained air.

9.7.7. Proximity of two gates

The proximity of two gates in a conduit can cause vibration of the downstream gate when the control gate is in an intermediate position and the guard gate is lowered. The jet from the guard gate gives rise to alternating forces on the control gate and in some combination of gate positions there can be turbulent recirculation of flow between the gates.

Nielson and Pickett[46] recorded vibration due to this cause and Petrikat[47] mentions a cure of a similar problem by jet dissipators which broke up the discharge jet from the guard gate. Naudascher[48] also draws attention to the danger of vibration due to two consecutively positioned gates in conduit. These are generally transitory problems; however, if considered acceptable they must still be of a level not to damage the structure.

9.7.8. Air demand

Flow below a gate

When a vertical-lift gate in a conduit is opened and the downstream section of the conduit contains no water, a demand for air arises due to entrainment of air in the issuing jet.

The total air demand consists of two different parts, air entrainment in the water flow as bubbles or larger air pockets in the air water transition region, and air flowing above the transition zone because of the drag of the flowing mixture. At initial gate openings, the issuing jet is accompanied by spray which entrains a high proportion of air.

To explain differences in air demand, flow has been classified[49]. The total air demand for free surface flow in conduits does not normally have its maximum at fully open gates. Often two maxima exist, one for very small gate openings, when spray flow occurs at 4–8% of gate opening, and a second one usually larger than the first when the gate opening is between 40–70%.

If a hydraulic jump occurs further air entrainment will occur. Kalinske and Robertson[50] expressed it in terms of the ratio of air flow to water flow. Air entrainment without jumps has been investigated by a number of researchers[49,51].

The suggested design assumption[52] is:

$$\beta = 0.03 \, (F_r - 1)^{1.06} \qquad\qquad 9.13$$

where β is the ratio of air flow to water flow
    F_r = Froude number $V/\sqrt{(gy)}$
    V = flow velocity at the vena contracta
    y = water depth at the vena contracta

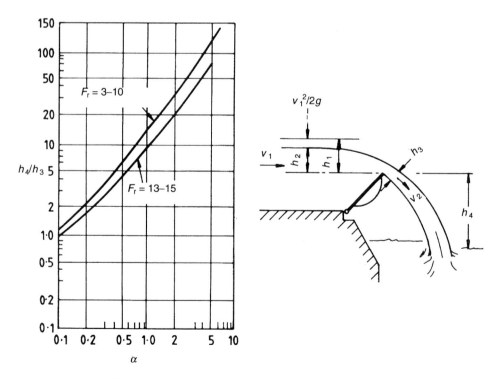

Figure 9.21. Coefficient of air demand for an overflow gate

The contraction coefficient for a gate with a 45° lip is 0.8. The suggested formula results in significantly more conservative values of air than that resulting from the investigation of Kalinske and Robertson[50].

The air admission pipes should be designed for velocities of not more than 40 m/s to prevent excessive pressure loss due to flow resistance in the ducts as well as entrance and exit air flow losses. These cause subatmospheric pressure conditions in the water conduit. Air flow losses can be calculated from the data in the CIBSE Guide[53].

Flow over a gate

The requirement to vent the nappe of an overflow gate was stated in Chapter 2. Air is entrained in the falling water and reduces the pressure under the gate unless it is vented. The subatmospheric conditions resulting from air evacuation cause the nappe to oscillate and the water level to fluctuate. This can result in severe gate vibration.

Venting is effected by flow dividers which break up the nappe locally and admit air through the openings created by divided flow. Ducts leading from pier level to the underside of a gate, Figure 2.24, are another method of venting. In most cases both means of admitting air have to be used together. In the extreme case the pressure under the gate can reach −10.33 m gauge. This would exert a negative pressure of 10.33 Mp/m² on the underside of the gate.

Air entrainment is proportional to the velocity head of the nappe. The air demand can be expressed as:

$$Q_A = \alpha Q$$

where Q_A = air demand
Q = flow over the gate
α = a coefficient depending on the height of fall of the nappe, h_4, the depth of the nappe, h_3, and Froude number of the nappe.

The depth of the nappe is approximately $0.6h_2$, where h_2 is the actual head above the gate lip. Therefore:

$$h_3 = 0.6\left(h_1 - \frac{v_1^2}{2g}\right)$$

where v_1 is the velocity of the approach flow to the gate.

The Froude number of the nappe is:

$$F_r = \frac{v_2^2}{gh_3}$$

where v_2 is the velocity of the overflow.

Figure 9.21 shows the coefficient of air demand against the ratio h_4/h_3 for two ranges of Froude numbers. The air ducts can be sized in similar manner to those required to satisfy the air demand for underflow gates. Since ducts for overflow gates are usually short compared with those for underflow gates, the duct entry and exit losses will be more significant when the friction loss of the air supply system is calculated. They should therefore be considered.

The ducts should be so arranged that the air supply in any position of overflow of the gate is not blocked by the downstream water level, and that the air is admitted under the gate. In order to achieve this, the outlets of the air supply ducts are staggered. It may even be necessary to stagger relative to one another the termination of the air supply pipes in opposite sluiceway walls. It is the usual practice to screen the outlets of the vent ducts. The screens must be set back from the face of the sluiceway so that they do not damage the side seal of the gate when it moves over the duct outlets.

References

1. Rouse, H (1949) editor: Engineering Hydraulics, *Proc. 4th Hydr. Conference*, Iowa Institute of Hydraulic Research, John Wiley.
2. Metzler, D E (1948): A Model Study of Tainter Gate Operation, MS Thesis, State University of Iowa, in *Proc. 4th Hydr. Conference*, Iowa Institute of Hydraulic Research, editor Rouse, H.
3. Lewin, J (1980): Hydraulic Gates, *Journ. I.W.E.S.*, 34, No.3, p.237.
4. Chow, V T (1959): *Open Channel Hydraulics*, McGraw-Hill.
5. Franke, P G; Valentin, F (1969): The Determination of Discharge Below Gates in Case of Variable Tailwater Conditions, *Journ. Hydr. Research*, Vol. 7, No.4.
6. Young, L R; Fellerman, L (1971): *Toome Sluices Calibration Tests*, B.H.R.A., Report RR 1105, Jul.
7. Buyalski, C P (1983): Canal Radial Gate Discharge, Algorithms and Their Use, *Proc. Speciality Conf. on advances in irrigation and drainage: Surviving External Pressures*, Jackson, USA, Jul, editors Borelli, J; Hasfurther, V R; Burman, R D, New York, USA, A.S.C.E., p.538–545.
8. Lewin, J (1983): Vibration of Hydraulic Gates, *Journ. I.W.E.S.*, 37, 165.
9. Vrijer, A (1979): Stability of Vertically Moveable Gates, *19th I.A.H.R. Congress, Karlsruhe*, paper C5.
10. US Corps of Engineers: *Tainter Gates in Open Channels—Discharge Coefficients (Free Flow)*, Hydraulic Design Criteria, sheets 320–4 to 320–7.
11. Toch, A (1952): *The Effect of a Lip Angle upon Flow Under a Tainter Gate*, Masters Thesis, State University of Iowa, Feb.
12. Gentilini, L B (1947): Flow Under Inclined or Radial Sluice Gates—Technical and Experimental Results, *La Houille Blanche*, Vol. 2, p.145.
13. US Corps of Engineers: *Tainter Gates in Open Channels—Discharge Coefficients (Submerged Flow)*, Hydraulic Design Criteria, sheets 320–8 to 320–8/1.
14. US Corps of Engineers: *Tainter Gates on Spillway Crests—Discharge Coefficients*, Hydraulic Design Criteria, sheets 311–1 to 311–5.
15. US Corps of Engineers: *Tainter Gates on Spillway Crests—Crest Pressures*, Hydraulic Design Criteria, sheets 311–6.
16. Milan, D; Habraken, P (1984): *Kotmale, Report on Spillway Radial Gate, Model Tests*, Neyrpic Thermohydraulic and Hydro-elasticity Laboratory. Not published.

17. US Corps of Engineers: *Gated Overflow Spillways, Pier Contraction Coefficients*, Hydraulic Design Criteria, sheets 111–5 & 111–6.
18. Naudascher, E: Locher, F A (1974): Flow-induced Forces on Protruding Walls, *Proc. A.S.C.E., Journ. Hydr. Div.*, Vol. 100, HY2, paper 10347, Feb.
19. White, F M (1979): *Fluid Mechanics*, McGraw-Hill, New York, USA.
20. Muskatirovic, J (1984): Analysis of Dynamic Pressures Acting on Overflow Gates, *I.A.H.R. Sym. on Scale Effects in Modelling Hydraulic Structures*, Essingen am Neckar, Germany, Sep.
21. Rogala, R; Winter, J (1985): Hydrodynamic Pressures Acting Upon Hinged-arc Gates, *Proc. A.S.C.E., Journ. Hydr. Engineering*, Vol. 111, No. 4, Apr.
22. Pethick, R W; Harrison, A J H (1981): The Theoretical Treatment of the Hydraulics of Rectangular Flap Gates, *19th I.A.H.R. Congress, Karlsruhe*, subject B (c), paper No. 12.
23. Naudascher, E; Rao, P V; Richter, A; Vargas, P; Wonik, G (1986): Prediction and Control of Downpull on Tunnel Gates, *Proc. A.S.C.E., Journ. Hydr. Engineering*, Vol. 112, No. 5, May.
24. Weaver, D S; Martin, W W (1980): Hydraulic Model Study for the Design of the Wreck Cove Control Gates, *Canadian Journ. of Civil Engineering*, Vol. 7 No. 2.
25. Thang, N D; Naudascher, E (1983): Approach-flow Effects on Downpull of Gates, *Proc. A.S.C.E., Journ. Hydr. Engineering*, Vol. 109, No. 11, Nov.
26. Kolkman, P A (1984): Phenomena of Self Excitation, in *Developments in Hydraulic Engineering—2*, editor Novak, P, Elsevier Applied Science Publishers.
27. Hite J E; Pickering, G A (1983): *Barkley Dam Spillway Tainter Gate and Emergency Bulkheads, Cumberland River, Kentucky; Hydraulic Model Investigation*, US Army Engineer Waterways Experiment Station, Vicksburg, Miss., Technical Report HL-83-12, Aug.
28. Grace, J L (1964): *Spillway for Typical Low-level Navigation Dam, Arkansas River, Arkansas; Hydraulic Model Investigation*, US Army Engineer Waterways Experiment Station, Vicksburg, Miss., Technical Report 2–655 Sep.
29. Daggett, L L & K G H (1974): Similitude in Free-surface Vortex Formations, *Proc. A.S.C.E., Journ. Hydr. Div.*, Vol. 100, Nov.
30. Anwar, H O; Weller, J A : Amphlett, M B (1978): Similarity of Free-vortex at Horizontal Intake, *Journ. of Hydr. Research*, Vol. 16, No. 2.
31. Gulliver, J S; Rindels, A J; Lindblom, K C (1986): Designing Intakes to Avoid Free-surface Vortices, *Water Power and Dam Construction*, Sep, p.24–28.
32. Gordon, J L (1970): Vortices at Intakes, *Water Power*, Apr.
33. Ables, J H (1979): *Vortex Problem at Intake Lower St Anthony Falls Lock and Dam, Mississippi River, Minneapolis, Minnesota*, US Army Engineer Waterways Experiment Station, USA, Technical Report HL-79-9, May.
34. Ziegler, E R (1976): *Hydraulic Model Vortex Study Grand Coulee Third Powerplant*, Engineering Research Centre, Bureau of Reclamation, Denver, Colorado, USA, Feb.
35. Song, C C S (1974): *Hydraulic Model Tests for Mayfield Power Plant*, University of Minnesota, St Anthony Falls Hydraulic Laboratory, Project Report No. 148, Apr.
36. Blaisdell, F W; Donnelly, C A (1958): *Hydraulics of Closed Conduit Spillways: Part X, the Hood Inlet*, Agricultural Research Service, St Anthony Falls Hydraulic Laboratory, Technical Paper 20, Series B.
37. Wagner, W E (1967): Glen Canyon Dam Diversion Tunnel Outlets, *Proc. A.S.C.E., Journ,. Hydr. Div.*, Vol. 93, HY6, Nov, p.113–134.
38. Ball, J W (1963): Construction Finishes and High-velocity Flow, *Proc. A.S.C.E., Journ. Construction* Div., Vol. 89, No. CO2.
39. Dickson, R S; Murley, K A (1983): Dartmouth Dam Low Level Outlet Aeration Ramps, *Ancold Magazine*.
40. Anastassi, G (1983): Besondere Aspekte der Gestaltung von Grundablassen in Stollen (Design of High-pressure Tunnel Outlets), *Wasserwirtschaft*, 73, 12.

41. Ball, J W (1959): Hydraulic Characteristics of Gate Slots, *Proc. A.S.C.E., Journ. Hydr. Div.*, Vol. 85, HY10, Oct, p.81–114.

42. Galperin, R (1971): Hydraulic Structures Operation under Cavitation Conditions, *14th I.A.H.R. Congress, Paris*, Proc. Vol. 5, p.45–48.

43. May, R W P (1987): *Cavitation in Hydraulic Structures: Occurrence and Prevention*, Hydraulic Research, Wallingford, Report SR79.

44. Koch, H J: Schußstrahlzusammenführung bei einem Grundablass mit Nebeneinanderliegenden Segmentschützen (Confluence of Two Jets Created by Two Parallel Segment Gates of a Bottom Outlet), *Wasserwirtschaft*, 72, 3.

45. Bruce, B A; Crow, D A (1984): *Mrica Hydroelectric Project: Hydraulic Model Study of the Culvert Control Structure*, B.H.R.A., report RR2325.

46. Nielson, F M; Pickett, E B (1979): Corps of Engineers Experiences with Flow-induced Vibrations, *19th I.A.H.R. Congress, Karlsruhe*, paper C3.

47. Petrikat, K (1979): Seal Vibration, *19th I.A.H.R. Congress, Karlsruhe*, paper C14.

48. Naudascher, E (1972): Entwurfskriterien für Schwingungssichere Talsperrenverschlüße (Design Criteria for Avoiding Vibration of High-head Gates), *Wasserwirtschaft*, 62, 112.

49. Sharma, H R (1973): *Air Demand for High-head Gated Conduits*, University of Trondheim, Oct.

50. Kalinske, F; Robertson, R A (1943): Closed Conduit Flow: Symposium on Entrainment of Air in Flowing Water, *A.S.C.E., Transactions*, paper No. 2205.

51. Wunderlich, W (1961): *Beitrag zur Belüftung des Abflusses in Tiefauslässen* (Commentary on Air Demand in Conduit Gates), Technische Hochschule, Karlsruhe.

52. US Corps of Engineers: *Air Demand, Regulated Outlet Works*, Hydraulic Design Criteria, Sheet 050–1.

53. Chartered Institute of Building Services Engineers: *Guide, C4–48 and C4–49*, Figure C4.3 Air flow in round ducts, Figure C4.4 Air flow in rectangular ducts.

10
Gate vibration

Gate vibration, when it occurs, can be a serious problem. It can result in structural damage or restrict operation at certain gate openings. In some cases vibration of a gate will occur at specific hydraulic conditions which may only occur years after commissioning of the installation. Even when these have been identified it may not be easy to reproduce them so that they can be investigated. Apparent steady state conditions may be subject to a minor hydraulic disturbance which overcomes the damping forces acting on the gate and initiates an unstable motion when oscillations occur with increasing amplitude. This chapter is intended as an introduction to the subject and offers some guidance on design features which will avoid vibration.

Many gates incorporate elements which are likely to result in vibration. In spite of this they have operated satisfactorily. One possibility is that disturbing forces are of low magnitude and are damped out. This can be the case with small gates in river courses. Also, since gates are designed for long return period events, the conditions which could cause vibration have not yet occurred. It does not follow that gates of similar design will be equally satisfactory at a higher head or scaled up in size. Most of the research papers on gates deal with vibration problems. This suggests that vibration is possibly the most frequent cause of malfunction of gates.

10.1. Types of gate vibration

Gate vibrations can be classified into three types[1]:

- Extraneously-induced excitation which is caused by a pulsation in flow or pressure which is not an intrinsic part of the vibration system (the gate).
- Instability-induced excitation which is brought about by an instability in the flow. Examples are vortex shedding from the lip of a gate and the alternating shear-layer reattachment underneath a gate.
- Movement-induced excitation of the vibrating structure. Under these conditions the flow will induce a force which tends to enhance the movement of the gate.

10.2. The vibrating system

The equation of motion of the simplest form of vibrating system with linear components (Figure 10.1) is given by:

$$m\frac{d^2y}{dt^2} + c\frac{dy}{dt} + ky = F(t) \qquad\qquad 10.1$$

where t = time
         m = mass
         k = spring rigidity
         c = damping (viscous)
         F = impressed force
         y = displacement

The total mass of a submerged gate or a gate under free discharge conditions is made up of its mass (m) and the hydrodynamic mass or added mass of water vibrating with the gate (m_w). Similarly there is a hydrodynamic component of damping (c_w) and rigidity (k_w).

For a system in water Equation 10.1 can be represented by:

$$(m+m_w)\frac{d^2y}{dt^2} + (c+c_w)\frac{dy}{dt} + (k+k_w)y = F \qquad\qquad 10.2$$

where F represents all hydrodynamic forces
         c_w = added mass damping
         k_w = added mass rigidity

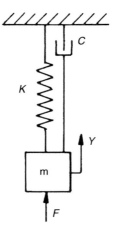

Figure 10.1. The vibrating system

The system is stable or positively damped when:

$$(c + c_w) > 0 \qquad\qquad 10.3$$

As a first approximation, if added mass damping is neglected:

$$c > 0 \qquad\qquad 10.4$$

The system is unstable or negatively damped when:

$$(c + c_w) < 0 \qquad\qquad 10.5$$

The critical damping coefficient C_c is given by:

$$C_c = 2m\omega_n = 2(m + m_w)\sqrt{\frac{k + k_w}{m + m_w}} \qquad\qquad 10.6$$

where ω_n = natural frequency of oscillation of the gate in water and the damping ratio ζ is:

$$\zeta = c/C_c$$

It now depends on whether the damping ratio ζ is greater or less than unity. When damping is less than critical $\zeta < 1$ and oscillation occurs with diminishing amplitude (stable or positively damped). If, as a first approximation, added damping c_w and added mass rigidity k_w are neglected:

$$\zeta = \frac{C}{2(m + m_w)\sqrt{[k/(m + m_w)]}} < 1 \qquad\qquad 10.7$$

Kolkman[2] gives an explanation of added damping and added rigidity. The added mass coefficient C_m is given by:

$$C_m = m_w/\rho D^2 L \qquad\qquad 10.8$$

where ρ = fluid density
D = gate depth or characteristic body dimension (gate immersion)
L = spanwise width of gate

In calculations, damping c is usually assumed to be constant friction damping. In slide gates the friction between the gate and the downstream bearing face and the seal friction provide damping. In roller gates it is the roller bearing friction, the rolling resistance as well as the seal friction. Where upstream reaction pads have been provided in order to eliminate transverse movement of the gate within the gate slots, they will contribute to the damping forces. However, the assumption of constant friction is not necessarily valid under conditions of gate vibration.

To prevent vibration the dominant excitation frequencies should be well away from resonance frequencies, given by the equation:

$$f_r = \frac{1}{2\pi} \sqrt{\frac{k + k_w}{m + m_w}} \qquad\qquad 10.9$$

The resonance frequency f_r must be a factor higher than the excitation frequency f due to the flow velocity or of a reflected pressure wave. Kolkman[3] suggests that one should try to obtain at least a factor 3. At the condition when the excitation frequency is equal or close to the resonance (natural) frequency, the displacement amplitude for the vibrating system increases very rapidly and may result in failure of the gate suspension system. The transmissibility rates, or the magnification factor T.R. is given by:

$$\text{T.R.} = \frac{1}{1 - (f / f_r)^2} \qquad\qquad 10.10$$

The transmissibility ratio should be negative to prevent excitation of the gate, that is the frequency ratio is greater than unity. The range between transmissibility ratios of unity and zero is sometimes called the isolation range with the percentage of isolation expressed between these limits. It is desirable to produce a design with a high percentage of isolation. With a frequency ratio of 3 recommended by Kolkman[3] the transmissibility ratio is 0.125 and the percentage isolation 87.5. At a ratio of 2, T.R. is 0.333 and at 1.5, T.R. is 0.800.

10.3. Excitation frequencies

Two possible sources of disturbing frequencies are the vortex trail shed from the bottom edge of a partly open gate and the pressure waves that travel upstream in a conduit to the reservoir and are reflected back. The frequency of a vortex trail, or in the case of any flow-induced vibration, can be defined by the Strouhal number:

$$S = f (L/V)$$

where f = excitation frequency

L = a representative length of the flow geometry (In the case of a tunnel gate, it is the width, or twice the projection of the gate into the conduit)

V = a representative flow velocity at the gate

$$V = \sqrt{(2gHe)}$$

where He = energy head at the bottom of the gate

The Strouhal number of a flat plate is approximately 1/7. The excitation frequency of a vortex shed from a gate may therefore be estimated as:

$$f = \sqrt{(2gHe)}/7L \qquad\qquad 10.11$$

The vortex trail will spring from the upstream edge of a flat bottom gate causing pressure pulsations at the bottom of the gate. Where the gate has a 45° lip or a larger angle, the vortex trail springs from the downstream edge, eliminating bottom pulsations.

Abelev[4,5] presented a number of studies to establish the dominant Strouhal numbers, S, and the excitation coefficient, C^1. The S values for horizontal excitation of a culvert gate are shown in Figure 10.2. For a flat-bottom gate S numbers for vertical excitation were given by Naudascher[6] and are reproduced in Figure 10.3.

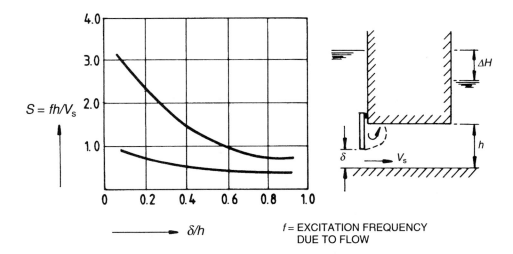

Figure 10.2. Strouhal number for the horizontal excitation of a culvert gate (after Abelev[4])

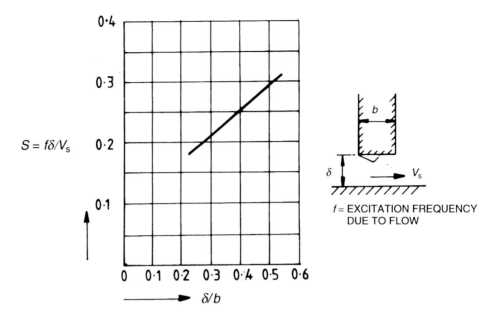

$S = f\delta/V_s$

Figure 10.3. Strouhal number for vertical excitation of a flat bottom gate (after Naudascher[6])

Only limited information is available on Strouhal numbers and a list for a wide variety of configurations is required to enable quantitative analysis of transmission ratios to be carried out for most gates.

The frequency of a reflected pressure wave is given by:

$$f = V\rho/4L \qquad\qquad 10.12$$

where $V\rho$ = velocity of the pressure wave (the value of $V\rho$ ranges from 1400 m/s for a relatively inelastic conduit to 1000 m/s for a relatively elastic pipe)

L = length of conduit upstream from the gate

The natural frequency of free vertical oscillation of a suspended gate is:

$$f_r = \frac{1}{2\pi}\sqrt{\frac{gE}{12s\sigma}} \qquad\qquad 10.13$$

where E = modulus of elasticity of the suspension (for rigid suspension rods, this is 200 kN/mm², but for wire ropes which stretch under load due to the spiral winding of the strand, E is usually taken as 70 kN/mm² (Table 7.2)

s = length of the suspension
σ = unit stress in the suspension (in calculating the unit stress, the mass of the gate and the added mass must be used $(m + m_w)$)

10.4. Added mass

Figure 10.4 illustrates the simplest case of added mass m_w. In this case, the added mass is the total mass of water above the piston and oscillation of the mass of the piston, m, forces the mass of water above to move with the same velocity. When a submerged body oscillates with small amplitude in a stagnant fluid, some of the fluid will oscillate in phase with the vibration of the body, but the further away the fluid is from the body the smaller its velocity compared with the body. The added mass m_w has the dimension of a virtual volume of fluid.

Computational methods have been established for determining m_w. They assume stagnant fluid conditions with potential flow induced by a body in harmonic vibration. The investigations were carried out by Wendel[7] Zienkiewicz and Nath[8] as well as Derunz and Geers[9].

Added mass coefficients C_m (see Equation 10.8) have been established experimentally by Hardwick[10], Hardwick *et al.*[11] and Thang[12]. The experimental work of Hardwick[10] and Thang[12] showed that the amplitude of vibration has little effect on C_m, but Thang found an appreciable effect due to frequencies between 9 to 12 Hz and some effect at vibration frequencies greater than 23 Hz (Figures 10.6, 10.7 and 10.8).

A gate stiffened by girders is analogous to the condition of Figure 10.4. The added mass and C_m will be considerably greater than that of a closed type of gate. The design of the gate bottom also has an appreciable effect on C_m. This is shown in Figure 10.9.

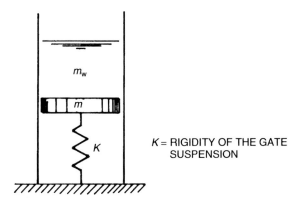

K = RIGIDITY OF THE GATE SUSPENSION

Figure 10.4. Added mass

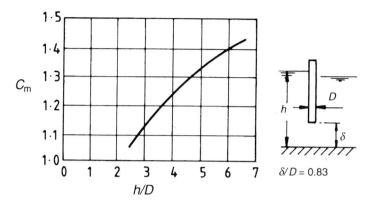

Figure 10.5. Variation of added mass coefficient with submergence for a flat bottom gate (after Hardwick[10])

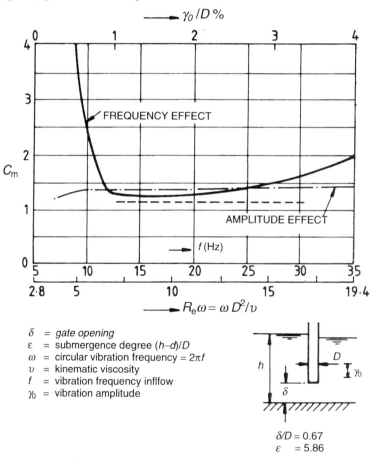

$$\delta = \text{gate opening}$$
$$\varepsilon = \text{submergence degree } (h-d)/D$$
$$\omega = \text{circular vibration frequency} = 2\pi f$$
$$\upsilon = \text{kinematic viscosity}$$
$$f = \text{vibration frequency inflfow}$$
$$\gamma_0 = \text{vibration amplitude}$$

$$\delta/D = 0.67$$
$$\varepsilon = 5.86$$

Figure 10.6. Effect of vibration behaviour on added mass coefficient C_m (after Thang[12])

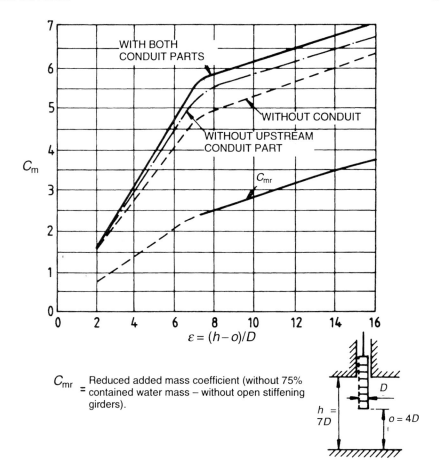

Figure 10.7. Effect of submergence and lateral confinement by gate slots and side walls on added mass coefficient C_m (after Thang[12])

10.5. Preliminary check on gate vibration

The sequence of carrying out a preliminary check whether the gate is liable to vibration is:

(i)   establish the total suspended mass (m)

(ii)   establish the spring rigidity of the suspension system (k)

(iii) determine the added mass (m_w).

(iv) calculate the damping forces due to slide or roller friction, roller bearing friction and seal friction. In all cases the sliding friction rather than the static friction should be taken, because a disturbance may initiate gate movement and the friction forces must then damp out the movement.

(v)   determine the critical damping coefficient C_c from Equation 10.6 and the damping ratio ζ.

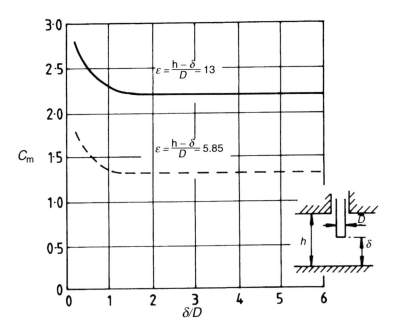

Figure 10.8. Effect of gate opening on added mass coefficient C_m (after Thang[12])

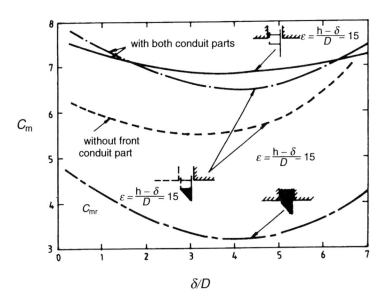

C_{mr} = REDUCED ADDED MASS COEFFICIENT
(WITHOUT 75% CONTAINED WATER MASS)

Figure 10.9. Effect of gate bottom (after Thang[12])

(vi)   calculate the resonance frequency f_r from Equation 10.9.

(vii) if the gate is in conduit and the gate lip is 45° or greater, calculate the frequency f of the reflected pressure wave from Equation 10.12.

(viii) establish the transmissibility ratio T.R. from Equation 10.10.

Other dynamic forces which can cause gate vibration are wave action, cavitation, two-phase flow and water column separation.

Kolkman[3] has suggested that vibration due to unsteady flow is probably due to a mechanism involving a fluctuation discharge coefficient, induced by the added mass flow of the vibrating gate. These conditions cannot be analysed theoretically and are described in subsequent sections of this chapter.

10.6. Vibration due to seal leakage

This is probably the most frequent cause of gate vibration. The mechanism of self-excitation due to seal leakage is explained by Petrikat[14] and Lewin[15]. The explanation of the hydrodynamic effect differs in the two papers. Petrikat gives an example of vibration due to the top seal for a low-level vertical-lift gate (Bharani Dam) and Krummet[16] discusses a similar example of vibration of a radial gate at a bottom outlet, due to leakage of the top seal. Other examples of vibration caused by seal leakage are given by Kolkman[3] and Mitchell[17].

10.6.1. Sill seals

Sill seals should be of rectangular shape and be moulded in a moderately hard elastomer (Shore A hardness 65) for gates in open channels and medium-head gates. For high-head gates a Shore hardness of 80 is more appropriate. Bottom seals can be metal to metal and this can overcome vibration induced by seals[18,19] (Chart D), [13] but perfect sealing with such an arrangement is difficult to effect with large and heavy gates[16] and sometimes also with smaller ones[18]. Under no circumstances should elastomeric seals project more than 5 mm below the faceplate of a gate and in high-head gates the projection should be no more than is required to effect a seal, which is about 3 mm. Wide block seals of timber or other materials are not suitable[19,20], because they can shift the point of flow attachment[21]. Musical note shape seals are also not suitable as sill seals[19(Chart D),14], although they appear to be in use, as evidenced from comparatively recent model investigations. Diaphragm seals on bottom-hinged gates are also vulnerable to vibration. A seal of this type failed on a bear-trap weir and was replaced by a sliding seal[22]. Figure 10.10 shows some arrangements of seals which are unsuitable, alongside the correct configuration.

Slight vibration initiated by sill seal leakage can usually be identified by vibration of skin panels and small amplitude ripples upstream of the gate.

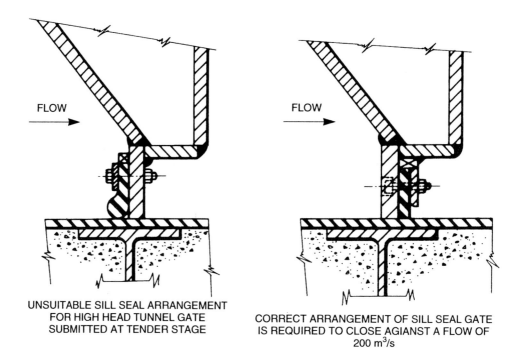

(a)

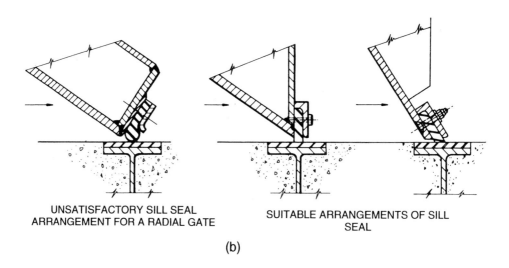

(b)

Figure 10.10. Arrangement of suitable and unsuitable seals

Severe vibration can cause high amplitude movement of a gate and be attended by loud noise.

10.6.2. Side seals

Leakage past the side seals of gates in open channels rarely causes vibration of a gate as a single unit but can initiate flexing of local structural members[23], sometimes severely[16,20]. It mostly results in seal flutter, which can be very noisy. Using two seals, one after the other, to suppress the jet from a leaking primary seal, is an unacceptable solution, because vibration can still be initiated by the primary seal. The junction between sill and lintel seals and of side seals, often presents design problems. Although special moulds are available for transition sections, these require rigid attachment. It is difficult to assess the incidence of vibration due to leakage at corners; it is probably high. The possibility of leakage at corners can be reduced by arranging side and horizontal seals in the same plane.

10.6.3. Lintel seals

Vibration of medium- and high-head gates due to flow past or impinging on lintel seals or protruding lips, was recorded by Petrikat[14,24] as well as Krummet[16]. Based on these and similar cases, centre-bulb seals should be used in preference to musical note shape seals (Figure 10.11). The flow past the opening—created at the lintel—once the seal is no longer in contact, can induce vibration[15].

In emergency closure gates and draft-tube inlet gates, when gates are used for initial filling of the tunnel, it is highly desirable (if not essential) for the seal to remain in contact with embedded parts of the lintel structure for the degree of opening required to fill the tunnel. Since narrow leak gaps lead to vibration[2], any design of an upstream sealing gate which aims at a rapid increase of the lintel seal gap after opening, will lead to a cantilever mounting of the seal[2] and be subject to excitation due to impingement of the flow.

10.7. Flow attachment, shifting of point of attachment and turbulent flow

10.7.1. Structural stiffening members

Structural stiffening members on radial and vertical-lift gates are frequently placed too low, so that intermittent flow reattachment occurs at the flange of the stiffening member. Figure 10.12(a) illustrates the design of the bottom section of a radial automatic gate which suffered severe vibration problems due to this cause. Figure 10.12(b) is a section through one of the Pershore Mill gates[21]—flat bottom gates, where flow attachment and vibration were predictable, as this shape is the most unstable because a free shear layer lies

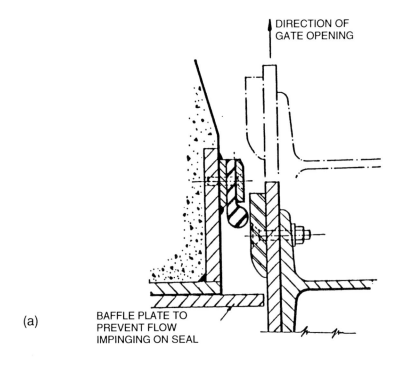

(a)

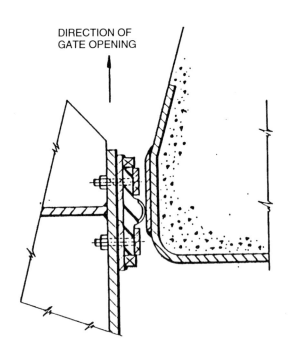

(b)

Figure 10.11. (a) A lintel seal arrangement which caused vibration before baffle plate was fitted[24]. (b) Preferred arrangement of lintel seal

close to the bottom of the gate[3]. Figure 10.12 (c) is the configuration of the bottom section of a diversion tunnel gate which would have led to flow reattachment problems because the structural stiffening member is placed too low.

10.7.2. Roller and turbulent flow downstream of a gate

When a gate operates under drowned-discharge conditions, an unsteady roller occurs and high turbulent conditions arise[26]. If the roller acts on submerged structural members, whether these are part of the gate skin plate stiffening or the arms of a radial gate, vibration is likely to occur. This is not so much a case of flow attachment as the hydrodynamic action of turbulent flow. Where structural members are submerged, the potential of flow-induced vibration is minimised when the member has a blunt trailing edge[27].

10.7.3. Gate design guide lines

In designing a gate the following guide lines can be stated:

- No structural member upstream or downstream of the control point should protrude into a line at 45° from the point of flow control, and upstream preferably at 60°[18] (Figure 10.13).
- It is better to arrange for vortex trail to be shed from the extreme downstream edge of a gate in order to achieve flow conditions that are as steady as possible[28].
- A sharp cut-off point should be provided at the lip[23,28].

Where a gate is situated near the crest of a weir and there is clear discharge downstream, hydrodynamic excitation does not arise either of gate arms or projecting structural stiffening members downstream of the skin plate. In tunnel gates, radial gates or vertical-lift gates in open channels, subject to downstream drowned conditions, hydraulic considerations should override structural priorities to dispose members in the most economical manner (Figure 10.14).

It appears that no investigation has been carried out of the effect on gates when the downstream discharge is in the region of an oscillating hydraulic jump. This is likely to cause problems where structural members are located low on the skin plate.

10.8. Hydraulic downpull forces and flow reattachment at the gate lip

Vibration due to elastic deflection caused by hydrodynamic downpull forces can occur in hydraulic servo-motor operated and in rope-suspended gates. In servo-motor operated gates, the problem can arise due to a long operating stem and the compressibility of the hydraulic fluid can also contribute to

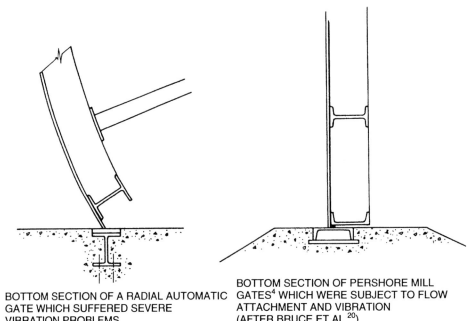

BOTTOM SECTION OF A RADIAL AUTOMATIC
GATE WHICH SUFFERED SEVERE
VIBRATION PROBLEMS

BOTTOM SECTION OF PERSHORE MILL
GATES[4] WHICH WERE SUBJECT TO FLOW
ATTACHMENT AND VIBRATION
(AFTER BRUCE ET AL.[20])

(a)   (b)

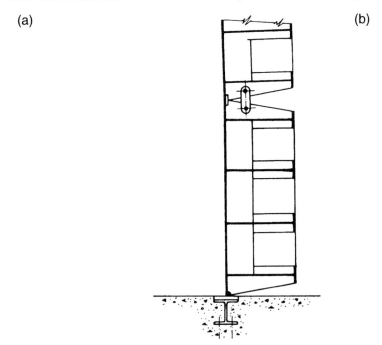

(c)   UNSUITABLE DESIGN OF THE BOTTOM SECTION OF
             A DIVERSION TUNNEL GATE. THE GATE IS
             REQUIRED TO CLOSE AGAINST A FLOW OF 150 m³/s.

*Figure 10.12. Arrangement of unsuitable structural stiffening members at the
bottom section of gates*

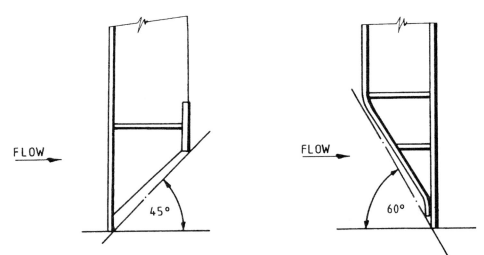

Figure 10.13. Arrangement of structural members at the bottom of a vertical lift gate

this. Elastic deflection due to ropes can be reduced by substituting chain suspension, but this can cause difficulties in the layout of the mechanical components where gates have to be hoisted through a considerable height. Chapter 9 gives some information on hydraulic downpull forces. In the absence of a hydraulic model study, a first approximation of investigating whether a gate is likely to vibrate due to hydraulic downpull forces can be made by following the procedure set out in Chapter 10 under 'Preliminary Check on Gate Vibration'.

Previous work suggests that the following guidelines will reduce downpull forces and conditions of separated flow and reattachment:

- Gate lips should have a sharp cut off point[18, 23, 28] (see also 10.7.3).
- Gate lips should be as narrow as possible.
- Gate lips should project as far as possible below the body of the gate[18, 25].
- The angle of inclination of the upstream face of the gate should be at least 45° (see also 10.7.3).
- The system should be as rigid as possible.

Various conditions of flow are illustrated in Figure 10.15 showing separated flow at the bottom of a gate and also the condition of possible shear layer deflection of entrapped fluid based on Martin *et al.*[29].

10.9. Unstable flow through small openings

Pressure fluctuations which, in turn, cause discharge fluctuations and thereby exert a force on the lower edge of a gate, can initiate gate vibrations at small openings. This is a self-exciting phenomenon which occurs at high-velocity flows under small gate openings or at leakage gaps at lintel seals when a gate is raised. It can be triggered by a vertical movement of the gate, which is then translated into a momentary pressure change that can reinforce the initial movement of the gate. The inertia of the water under flow conditions contributes by causing a pressure rise in the conduit. The gate movement will then be amplified under resonance conditions or damped out.

Vibration at small gate openings has been investigated by Kolkman[2,3] as well as Vrijer[28]. Kolkman suggests that the width of the leakage gap should be at least 1.5 times (preferably 2 times or more) the width of the gate edge. This criterion should also be used for minimum gate openings of tunnel gates for cracking open to fill downstream sections of the tunnel. If the resultant flow is unacceptably high as a consequence, a bypass system for filling should be provided. Experience suggests that gates will often remain stable at gate openings below the minimum recommended values, probably because the damping forces have been under-estimated. In Section 9.7.2 it was stated that minimum gate opening should not be less than 100 mm. If Kolkman's criterion for minimum leakage gap results in a greater opening, this should be considered the minimum.

10.10. Flow over and under the leaves of a gate

Instability due to vortex trails and/or flow separation where a gate leaf or a stoplog has to be lowered into flowing water has been investigated by Brown[30] and Grzywienski[31]. Vortex action is a real threat if the depth of a flow over the gate exceeds 0.45 of the gate height for a flat sharp-crested gate leaf. The depth of underflow is less critical. When sections of the gate or stoplogs can be connected together to avoid complete immersion, instability can be eliminated. Excitation can be disturbed by introducing another jet into the wake, for instance, through a controlled opening in the skin plate.

10.11. Vibration of overflow gates due to inadequate venting

The means to effect venting of the nappe of overflow gates was dealt with in Chapter 2. The underside of the nappe entrains air and causes sub-atmospheric pressure in the absence of venting or an inadequate air supply. If the air is exhausted it results in collapse of the nappe when it will suddenly attach itself to the underside of the gate with a violent impact.

In an inadequately vented overflow gate, the level of water on the inside of the nappe fluctuates due to inertia, causing variations in pressure which result in horizontal movements of the nappe. These in turn cause fluctuations of discharge, since a reduction in pressure increases the discharge over the gate. This sequence of events can result in severe gate vibration which may be transmitted to the civil engineering structure.

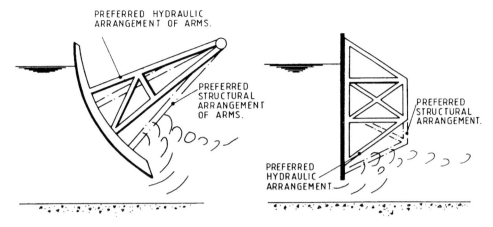

Figure 10.14. Preferred hydraulic structural arrangement of a gate subject to drowned discharge

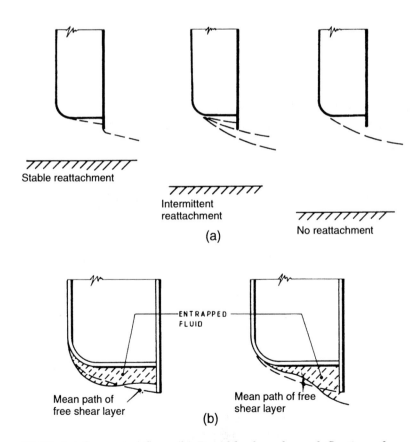

Figure 10.15. (a) Separated flow. (b) Possible shear layer deflection of entrapped fluid

10.12. Vibration due to a free shear layer

In addition to the possible shear layer deflection of water trapped under a gate as shown in Figure 10.15, vibration due to a free shear layer has occurred.

The model of the Split Yard Creek Control Structure in Queensland, Australia[32] indicated a well-defined shear layer at the gate shaft opening. This was subject to apparently periodic oscillations. Two control gates were located upstream of the shaft and in the fully open position were subject to flow-induced excitation which appeared as beats because one frequency component of the excitation was close to the natural frequency of the gate assembly. The problem was cured by a combination of leading- and trailing-edge ramps at the intersection of the shaft and the tunnel.

10.13. Two-phase flow

Where air can be introduced into a conduit, severe pressure fluctuations can occur at the control gate due to the build up of stagnated air under high pressure at the conduit crown upstream of the gate. The air, which is uniformly distributed at the head race tunnel, accumulates to form air pockets due to the relatively low velocity of flow in the conduit and the long distance upstream from the gate. The air pockets stagnate at the upstream side of the skin plate until they are partially drawn under the gate. When the pressurised air is released, it reaches atmospheric value almost instantaneously with explosive force.

A particularly severe problem of this type was noted in a model study by Rouvé and Traut[33] (Figure 10.16). In the discussion of the paper Kenn pointed out that air-entraining water flows are notoriously difficult to model, except perhaps when tested with full-scale velocities. Because of scaling problems, pressure fluctuations in prototypes may prove less severe than those suggested by model tests.

Nielson and Pickett[19] record severe vibration of a reverse radial gate which was attributed to the collapse of large vapour cavities near the gate. The gate acted as a control valve for a high-lift lock with a maximum differential head of 28.1 m. This type of problem can be solved only by venting upstream as well as downstream of the gate.

Singh *et al.*[34] reported the dislocation of a bulkhead on a tower type intake due to air compression. Air-entraining vortices had formed at the intake under some lower reservoir levels. The subsequent operation of the emergency gate 200 m downstream of the portal caused the surges which increased the pressure on the trapped air, driving it up the intake in an air/water spout which dislocated the bulkhead.

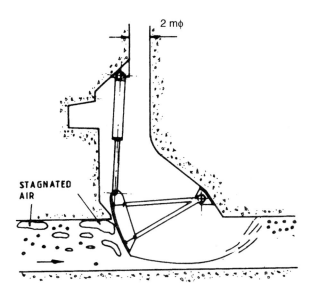

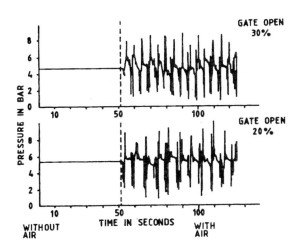

Figure 10.16. Two-phase flow below a radial gate

10.14. Slack in gate components

When a gate opens or closes, the inertia of the water creates regions with an increase or decrease in pressure. Excitation can also be brought about by flow velocity fluctuations at constant gate openings which will vary across the opening.

Vertical-lift roller gates can be subject to hydrodynamic pressure conditions so that the uppermost wheel or wheels are on the point of unloading.

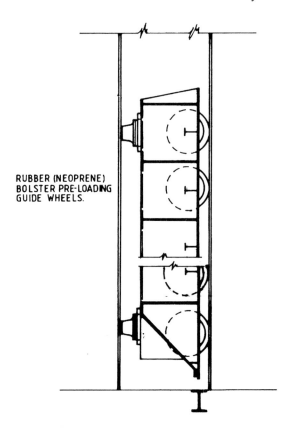

Figure 10.17. Pre-loading of the guide wheels of a vertical lift gate

Kolkman[3] gives an example of this type where a strong rotational vibration occurred, centred on the lower wheel shaft.

Excessive slack in mechanical gate components such as guide wheels, pivots or hoist chains should be avoided. Where clearances are essential it is important that the component be pre-loaded. An example of pre-loading of the guide wheels of a surge shaft gate is shown in Figure 10.17.

References

1. Naudascher, E (1979): On Identification and Preliminary Assessment of Sources of Flow Induced Vibration, *19th I.A.H.R. Congress, Karlsruhe,* paper C1.
2. Kolkman, P A (1984): Vibration of Hydraulic Structures in *Developments in Hydraulic Engineering—2*, editor Novak, P, Elsevier Applied Science Publishers.
3. Kolkman, P A (1979): *Development of Vibration Free Gate Design*, Delft Hydraulics Laboratory, Publ. 219.
4. Abelev, A S (1979): Investigation of the Total Pulsating Hydrodynamic Load Acting on Bottom Outlet Sliding Gates and its Scale Modelling, *8th I.A.H.R. Congress, Montreal,* paper 10A1.

5. Abelev, A S (1963): Pulsations of Hydrodynamic Loads Acting on Bottom Gates of Hydraulic Structures and their Calculating Methods, *10th I.A.H.R. Congress, London*, paper 3.21.

6. Naudascher, E (1964): Hydrodynamische und Hydro-elastische Beanspruchung von Tiefschützen, *Der Stahlbau*, No. 7 and 9.

7. Wendel, K (1950): Hydrodynamische Massen und Hydrodynamische Massentraggeheits-momente, *Jahrbuch der Schiffsbautechnischer Gesellschaft*, 44, 207–55.

8. Zienkiewicz, O C; Nath, B (1964): Analogue Procedure for Determination of Virtual Mass, *Proc. A.S.C.E., Journ. Hydr. Div.*, HY5, Sep, p.69.

9. Derunz, J A; Geers, T L (1978): Added Mass Computation by the Boundary Integral Method, *Int. Journ. Numerical Methods* Eng,12, 531–49.

10. Hardwick, J D (1969): *Periodic Vibrations in Model Sluice Gates*, Ph.D. Thesis, Imperial College of Science and Technology, London.

11. Hardwick, J D; Ken, M J; Mee, W T (1979): Gate Vibration at El Chocon Hydro-power Scheme, Argentina, *19th I.A.H.R. Congress, Karlsruhe*, paper C7.

12. Thang, N D (1982): Added Mass Behaviour and its Characteristics at Sluice Gates, *Int. Conf. On Flow Induced Vibrations in Fluid Engineering*, B.H.R.A., Reading, England, Sept, paper A2.

13. U.S. Waterways Experimental Station (1956): *Vibration and Pressure Cell Tests, Flood Control Intake Gates Fort Randall Dam, Missouri River*, South Dakota, Technical Report No. 2–435, Vicksburg, Mississippi, Jun.

14. Petrikat, K (1979): Seal Vibration, *19th I.A.H.R. Congress, Karlsruhe*, paper C14.

15. Lewin, J (1983): Vibration of Hydraulic Gates, *I.W.E.S.*, 37, 165.

16. Krummet, R (1965): Swingungsverhalten von Verschlussorganen im Stahlwasserbau, *Forschung in Ingenieurwesen*, Bd. 31, No. 5.

17. Mitchell, W R (1979): Vibration Due to Leakage Through a Reverse Radial Gate, *19th I.A.H.R. Congress, Karlsruhe*, paper C17.

18. Hart, E D; Hite, J E (1979): Barkley Dam Gate Vibrations, *19th I.A.H.R. Congress, Karlsruhe*, paper C15 (Charts B and E).

19. Nielson, F M; Pickett, E B (1979): Corps of Engineers Experience with Flow Induced Vibrations, *19th I.A.H.R. Congress, Karlsruhe*, paper C3.

20. Lewin, J (1980): Hydraulic Gates, *I.W.E.S.*, 34, No. 2, p.237.

21. Bruce, B A; Crow, D A (1978): *Hydroelastic Model Studies of the Pershore Mill Sluice Gates*, B.H.R.A., Report RR 1485, Jul.

22. Merkle, T (1979): Hydraulically Induced Vibrations in a Bear-trap Weir, *19th I.A.H.R. Congress, Karlsruhe*, paper C19.

23. Schmidgall, T (1972): Spillway Gate Vibrations on Arkansas River Dams, *Proc. A.S.C.E., Journ. Hydr. Div.*, HY1.

24. Petrikat, K (1976): Structure Vibrations of Segment Gates, *8th I.A.H.R. Congress, Leningrad*.

25. Hampton, I G; Lesleighter, E J (1980): Effect of Gate Shape on Closure Loading, *7th Australasian Hyd. and Fluid Mechanics Conference, Brisbane*, Aug.

26. Murphy, T E (1963): Model and Prototype Observations of Gate Oscillations, *10th I.A.H.R. Congress, London*.

27. Pennino, B J (1981): Prediction of Flow Induced Forces and Vibration, *Water Power and Dam Construction*, Feb, p.19.

28. Vrijer, A (1979): Stability of Vertically Movable Gates, *19th I.A.H.R. Congress, Karlsruhe*, paper C5.

29. Martin, W W; Naudascher, E; Padmanabham, M (1975): Fluid Dynamic Excitation, Involving Flow Instability, *Proc. A.S.C.E., Journ. Hydr. Div.*, HY6, Jun.

30. Brown, F R (1961): Fluctuation of Control Gates, *9th I.A.H.R. Congress, Dubrovnik*, p.258–269.

31. Grzywienski, A (1963): The Effect of Turbulent Flow on Multi-section Vertical-lift Gates, *10th I.A.H.R. Congress, London*.

32. Chang, H T; Hampton, I G (1980): Experiences in Flow Induced Gate Vibrations, *Int. Conference on Water Resources Development*, Taipei, Taiwan, May.

33. Rouvé, G; Traut, F J (1979): Vibrations Due to Two-phase Flow Below a Tainter Gate, *19th I.A.H.R. Congress, Karlslruhe*, paper C10.

34. Singh, S; Sakhuja, V S; Paul, T C (1982): Some Lessons from Hydro and Aero-elastic Vibrations Problems, *Int. Conference on Flow Induced Vibrations in Fluid Engineering*, Reading, Sept, paper B1.

11
Control systems and operation

This chapter deals with control objectives, operating rules and systems, telemetry, fall-back systems and standby facilities, as well as instrumentation. In control systems, the trend is for electro-mechanical controls to be superseded by closed-loop control systems (electronic controls). A contributory factor is the increasing complexity of operating rules which may require flood attenuation in the discharge from a spillway to a river course, or in barrages the gates may have to be operated in a different manner when dealing with a flood compared with the incidence of high tides or storm surges. Appropriate responses can be programmed into electronic controllers.

Where different responses to different conditions are required, operator training and practice present problems because there is little or no experience of rare or extreme events. Training may have to be carried out on a simulator.

Invariably automatic gate control systems are backed by manually-operated standby controls. Safety and reliability engineering requires that operating and standby systems are independent of one another and are functionally different. The practice is for an electronic controller to be backed by a second electronic controller and for automatic change over from one controller to the other in the event of malfunction.

11.1. Control objectives

11.1.1. In rivers

The operation of gates in rivers is to maintain the upstream water level for navigation or for water abstraction and to pass flood flow. Gates may be used to lower the water level for construction purposes, such as bank consolidation, or for channel improvement works. Such operations are carried out under the control of an operator. Automatic control is mainly based on upstream level control. Where flood warning systems are in operation the water level in a reach or in a reservoir may be lowered in anticipation of a flood.

11.1.2. In reservoirs

Safety of the dam

This is ensured by preventing the reservoir level from rising to within a few metres of the crest as overtopping could destroy fill dams or other types of dam.

Maximum storage

For hydro-electric and for many irrigation projects it is required that the maximum operating level should be reached at the end of the flood to maximise generation and storage for irrigation release.

Flood routing

Some of the following requirements may apply under conditions of flood routing[1,2]:

• The maximum flood discharge must not be increased and preferably should be significantly attenuated (Figure 11.1).
• The rise in the rate of flow must not be increased.
• The flood propagation rate should be attenuated (Figure 11.2).
• The provision of storage capacity in expectation of a flood or snowmelt.
• Bank Stability—in the reservoir by avoiding rapid rise or fall of water level
                    —downstream of the reservoir by limiting channel velocities.

11.2. Operating rules and systems (manual methods)

Local control of motorised operation

This takes the form of increasing the outflow in steps. At its simplest level the gate openings may be determined by the operator from a signal of river or reservoir level and be based on his experience.

If it is required to reduce operator judgement, curves are used which show in 30–60 minute intervals the increase in gate opening as a function of river or reservoir level and rate of change of level during the interval[3]. The charts may be designed so that, within a range of the rate of change of level, a specified discharge rate is not exceeded in order to limit downstream flooding.

Remote control of motorised operation

This may be located remote from the spillway, barrage or weir or from a centre controlling a number of dams. In all cases alternative or standby operation of the gates is provided at the spillway or barrage.

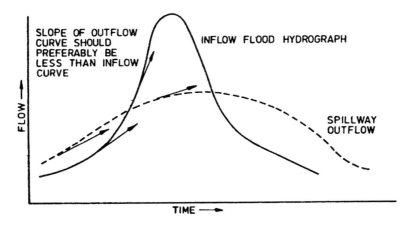

Figure 11.1. Attenuation of outflow and reduction in the rate of outflow compared with inflow

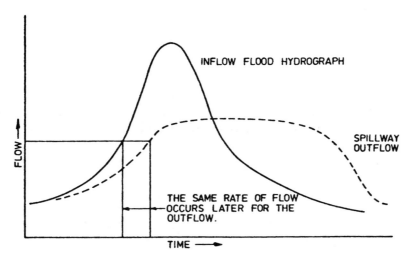

Figure 11.2. Attenuation of flood propagation rate

Computer-assisted control

Extensive input data may require the use of a computer to determine how the gates should be operated. This is the case when a hydro-meteorological model is used, or when the complexity of interpreting operation charts is likely to lead to errors.

For example, data from raingauge stations, river gauging, inflow into the reservoir, reservoir level, meteorological information and gate opening may be fed into a computer, which then, by reference to the available flood storage, carries out the flood routing calculations and prints out the operating instructions. These are then executed by the personnel (River

Medway flood prevention scheme[4]). The data may be entered manually into the computer, or the computer may operate on signals received from the gauging stations and control instruments.

11.3. Operating rules and systems (automatic methods)

Cascade controls

Cascade control is mostly used in the control of reservoirs and its application will therefore be discussed in that context. The distance between the retention level and the maximum reservoir level is divided into a series of steps, each corresponding to a gate opening. On reaching a specific water level, the gate hoist motion is started and the gates open in sequence to their predetermined height, controlled by limit switches. A frequent refinement is to provide alternative limit switches, permitting greater gate openings if one of several spillway gates is out of operation due to maintenance or malfunction.

Cascade control is generally used in conjunction with power actuation of gates, either by electric motor driven winches or by oil hydraulic cylinders. At the Victoria Dam in Sri Lanka counterbalanced radial spillway gates were devised[5], which can open under gravity. The control system is of the cascade type but can operate mechanically without external power, with electrical controls as a standby. Closure of the gates is by oil hydraulic cylinders supplied by electric motor driven, oil hydraulic power packs.

There are eight gates, each 12.5 m wide, which open in pairs in four stages at 0.7, 2.5, 4.7 and 9.35 m. Opening is actuated by floats, which over a rise of upstream water level of 0.64 m operate oil hydraulic poppet valves. These direct the oil from the piston side of the hoist cylinders to the tank, permitting the gates to open under gravity. When the gates have reached their appropriate opening step, an actuator on the gate closes another poppet valve which stops the flow of oil and locks the cylinder in position. Other floats are connected over pulleys to electric limit switches which energise the relays for solenoid operation of oil hydraulic, directional control valves. This provides a standby system for opening the gates.

Electro-mechanical level control

This is actuated by a predetermined rise in the water level above the retention level, which initiates opening of the gates in steps and in sequence. A water-level control band is set. When the upper limit of the control band is reached, the opening motion is started and continues in steps until the level falls below the upper limit, when the motion stops. Closure of the gates commences when the lower limit of the control band is reached; the raising and lowering motions are interrupted by a dwell period to prevent hunting. An ultimate upper water-level limit switch initiates an alarm signal and some control systems are designed so that the gate hoist dwell period is cancelled during the time when the uppermost level is reached or exceeded.

Level control by computer (feed-back control system[6])

This moves the gates in turn to a set point after determining the desired outflow with reference to the measured inflow. The control instructions are given by a computer which is programmed with the strategy for maintaining upstream water level.

In proportional integral derivative control (PID control) (see also p. 200) the value of the upstream water level to be maintained is compared with the value transmitted by water level sensors. The difference between the two signals, the error, is computed at fixed intervals and is used to operate the proportion and the integration algorithm. The proportional term causes an immediate and longer-term corrective action, and as long as the error persists the integral term will increase or decrease continuously so as to open or close the spillway gates.

Feed-back level control systems[6] compare the actual value of a variable with its desired value and take the necessary corrective action. In gate control the variable is usually the water level, although the rate of change of water level may also be used as an additional control parameter. The system characteristics of a good closed-loop system based on feed back, operate to maintain the desired level, the set point, to correct for any variations with a minimum of oscillation. The gate should respond so that a small error results in a small opening and a larger error in a larger gate opening. The actual as compared with the desired level should be closely tracked by the system.

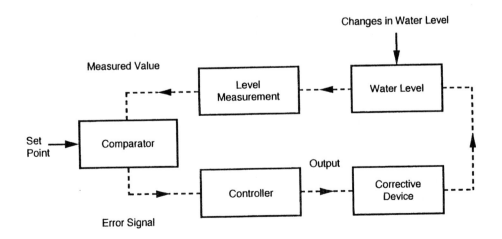

Figure 11.3. Block diagram of a closed loop system

Closed-loop control system

The system should be judged on the following criteria:

- How well the system reduces the error signal to zero or almost zero.
- The final difference between the measured value and the set point. This is called 'offset' in control terminology.
- The speed with which a system responds or restores agreement. In gate operation speed of response is not an important factor.
- The system should be free of large and violent oscillations — that is, it should be stable.

Modes of control

There are four modes of control:

- Proportional control — magnitude oriented.
- Proportional plus integral (PI) control — magnitude and error time duration oriented.
- Proportional plus derivative (PD) control — magnitude and error rate of change oriented.
- Proportional plus integral plus derivative (PID) control — magnitude, error time duration and error rate of change oriented.

Proportional control

In proportional control the output is proportional to the error signal. When there is an increase in water level due to increased inflow the controller will actuate the raising of the gate to compensate for the increase in flow. Since the gate opening is proportional to the error signal, the new opening can only be maintained if there is a permanent error, therefore proportional systems tend to have a permanent error, the offset.

To overcome the offset problem the time integral to the error signal (magnitude of the error signal multiplied by the duration of the error) is used to determine the new gate opening. The proportional term positions the gate in proportion to the error signal, i.e. the increase in water level, and the integral term senses the offset which remains and continues the motion in the same direction until the offset is reduced.(Figure 11.4).

Proportional plus integral plus derivative control (PID)

The derivative term measures the rate of change and causes the system to react more rapidly (Figure 11.5). PID control is most frequently used in position control.

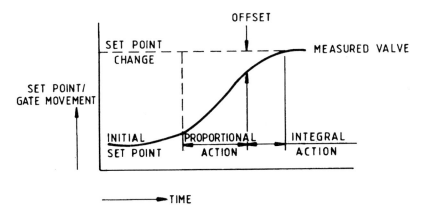

Figure 11.4. Proportional plus integral control

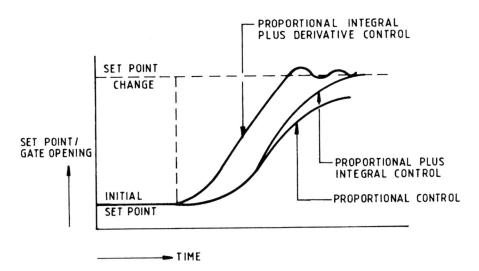

Figure 11.5. Comparison of different control modes

Application of PID control

The following symbols are used:

x_1 = required upstream water level
x_2 = actual water level
x_e = $x_2 - x_2$ = error signal
T_s = sampling interval
V_p = output of proportional term algorithm
V_i = output of integral term algorithm
V_t = $V_p + V_i$
q = nominal demanded flow rate

θ = gate angle, which is corrected to be proportional to gate opening

T_{gs} = Interval between gate error signals

K_1 = Proportional constant

K_2 = Integral constant

n = Level tolerance (of water level), also known as dead band

A = Surface area of reservoir or reach of river

The value of the upstream water level to be maintained x_1, is compared with the value transmitted by a water level sensor x_2. The sensor may be a float, an electrode, an ultrasonic device, a pressure transducer or a bubbler device.

The difference between the two signals $x_e = x_1 - x_2$ is computed every T_s seconds and is used to operate the proportion plus integration algorithm. The output V_t of this algorithm is:

$$\begin{aligned} V_p &= K_1 x_e & \text{the proportional term} \\ V_i &= V_p + K_2 x_e T_s & \text{the integral term} \\ \text{and} \qquad V_t &= V_p + V_i \end{aligned}$$

An upper and lower limit is placed on V_i (maximum and zero flow respectively).

The difference between the retention level and the actual water level (x_e) requires that the gate rises to discharge the increased inflow into the reach or the reservoir and a level difference (x_e) will correspond to a specific gate opening. The gate angle is not of course directly related to gate opening, and the coefficient of discharge for flow under a gate varies with gate opening. The computer has therefore to be programmed to convert the required flow rate into a set of demanded gate angles using a polynomial fit. This may require inbuilt logical hysteresis. The computer determines from the signal V_t the required outflow rate and converts it to the new gate angle θ.

For free-discharge conditions, the upstream water level and gate angle are input to the controller. For gates in barrages where the discharge can be submerged, the downstream water level has also to be measured and transmitted to the computer to enable the required outflow to be computed and converted to demanded gate angle. From gate control consideration where upstream level has to be maintained, the exact relationship between gate angle, i.e. gate opening and discharge under or over a gate, is not essential. It is only required if the data are logged to obtain a record of the flood hydrograph.

In multi-gate installations, each sluiceway will normally behave as if it is independent of the adjoining sluiceways, provided the approach to the gates is sensibly straight. Therefore the computer can be programmed to calculate θ the gate opening, depending on how many gates are operational.

The time taken to compute the gate angle is very small so that the value of the demanded gate angles will now be held for the rest of the main sampling period T_s. The demand angles are compared with the measured angles of the gates and form gate error signals every T_{gs} seconds. T_{gs} must be considerably shorter than T_s because the dynamics of the gate loop are very much faster than the response of the reservoir or river reach.

If any of the error signals exceed a threshold, the gate hoist motors are started up and close or open the gates to adjust the flow. This adjusts the upstream water level according to the difference.

During a flood, the inflow will rise and subsequently fall. Any difference between the demanded level and the measured level will:

• Cause an immediate corrective action due to the proportional term
• Cause a longer-term corrective action to eliminate any difference between x_1 and x_2 (the error).

The integral term operates so that if there is any difference between x_1 and x_2 (the error), the term will increase or decrease continuously causing the gates to open or close.

Hence there is no requirement within the control system to have an accurate relationship between the actual and the demanded outflow rates. If there is inconsistency between the two, the integral action will compensate for it. Ultimately, after a change of inflow to the reservoir or the reach, the flow over or under the gates will balance to eliminate any difference between x_1 and x_2.

In order to design the control system and the parameters T_s, T_{gs}, K_1 and K_2, the following variables must be fixed:

• The maximum allowable change in water level above retention for a given per centage change in the inflow and the time at which this maximum should occur (Figure 11.6).
• The time for the transient to settle within a specified proportion of the peak (Figure 11.6).
• The maximum permitted variation in level (shown in Figure 11.6). The time at which it should occur, together with a time at which this should decay to a given tolerance band (limits indicated in Figure 11.7).

In practice this is sometimes approached differently. When a gate is operated by an electro-mechanical hoist, it is advisable to limit the number of motor starts per hour to ensure long life. If, for instance, it is fixed at four starts per hour, a check is required that gate opening is compatible with the rate of change of upstream level at the steepest part of the design hydrograph, and that the change in water level above retention during the rest period of the motor is acceptable. This will usually be the case when all gates in a multi-gate sluice installation are in operation. It may be critical when one gate out of a two- or three-gate sluice is out of action due to maintenance or a defect. The gate controller can be programmed to override

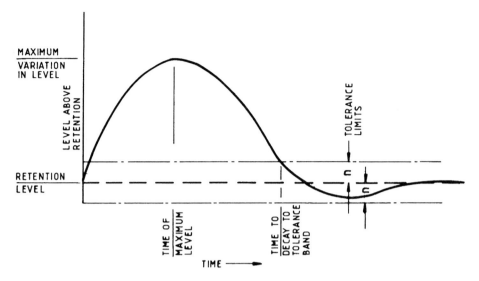

Figure 11.6. Illustration of upstream level variation due to increased inflow

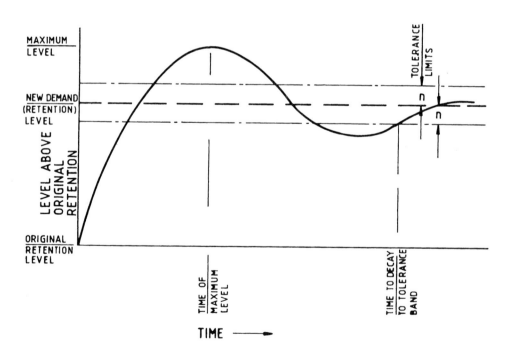

Figure 11.7. Illustration of upstream level variation due to a change in demand level

any restriction on the frequency of operation when one or several gates are not available, or when the increase in upstream water level exceeds a critical value.

When gates are operated by oil hydraulic servo-motors, frequency of motor starting is less important because motors driving oil hydraulic pumps are started with the pump off-loaded.

Any one operation of a gate installation must be such as to minimise surges upstream and downstream. This could be important to navigation, fishermen and other river users. It could also be important where there is a sluice installation under automatic control in the downstream reach. Where this possibility exists, transients in upstream level measurement must be filtered out. The usually adopted gate hoisting rate of 300 mm per minute does not prevent surges in rivers unless the hoisting time is limited.

For a given change in flow rate through the sluices, or a given change in demanded level, there will be a time limit for the system to respond, depending on the flow discharge due to the gate opening and the upstream head or the difference between upstream and downstream head. The speed of response is also influenced by the rate of inflow. To a first approximation, the rate of fall in level is given by:

$$\text{Rate of fall in level} = \frac{\text{outflow rate} - \text{inflow rate}}{\text{surface area of reservoir or reach}} = \frac{q_2 - q_1}{A}$$

The speed of response can be made higher for small signals by increasing K_1 and K_2. Making K_2 larger makes the system faster but more undamped, i.e. more and more oscillatory with larger and larger overshoots and undershoots.

A usual and conservative limit for a maximum outflow of q_2 is:

$$K_1 n = \frac{q_2 \max}{100}$$

where n = level tolerance

For the Kotmale reservoir[7,8]:

$$
\begin{aligned}
q_2 \max &= 5560 \text{ m}^3/\text{s} \quad \text{(the probable maximum flood)} \\
A &= 6.63 \text{ km}^2 \quad \text{(surface area of the reservoir)} \\
n &= \text{say 20 mm} \quad \text{(level tolerance)} \\
\text{then} \quad K_1 &= 2780 \text{ m}^2/\text{s}
\end{aligned}
$$

and in order to obtain adequate damping of the oscillations, it is advisable to place a limit on K_2 given by:

$$K_2 \leq K_1^2/A$$

For Kotmale, this would result in:

$$K_2 = 1.16 \text{ m}^2/\text{s}^2$$

The inter-sample interval T_s must be short enough not to destabilise the system. It is suggested that there should be at least 20 samples per cycle of transient oscillation. The cycle time T of the system will be of the order of:

$$T = 2 \pi \sqrt{(A/K_2)}$$

Hence for Kotmale:

$$T \cong 15021 \text{ s}$$

Therefore it is inadvisable to use an inter-sample interval T_s greater than $15021/20 \text{ s} = 750 \text{ s}$.

Other control parameters, apart from upstream level can be used, such as constant downstream level, which is common for irrigation channels or less frequently constant discharge.

11.4. Telemetry

Automatically-operated sluices and spillway gates, in most cases, are remotely supervised by telemetry at a central monitoring station. Usually the station supervises several installations, sometimes carrying out different functions. Key information is displayed on a VDU and recorded on an event printer. For a sluice gate installation, in addition to essential data, such as upstream and downstream water levels and gate positions, other data will be transmitted such as:

* Availability of mains supply.
* Availability of transformers.
* Condition and availability of standby generating plant.
* Gate availability.
* Aggregated faults in any one motor circuit or gate equipment.
* Valve position (if valves are part of the installation).

If the electrical switchgear and standby generators are located in a control building, other data will be transmitted such as:

* Intruder alarm.
* Fire alarm.

In general, telemetry transmission is confined to information and data transmittal. If these indicate a malfunction or a potentially dangerous condition, personnel are sent to the site to deal with it. In a few cases, instructions and commands are sent to the gate installation by telemetry.

Where operating control or overriding control of an automatic system is via telemetry, such as in a large barrage, the telemetry lines are duplicated. High reliability requires that an emergency control system is independent and generically different from the primary controls. This is effected by hard wiring the emergency controls and arranging for this method of operating the gates and associated plant to be located in a room different from that of the primary control system. This avoids a common cause fault, such as a fire, putting the installation out of operation.

Where the gate installation and the supervisory station are some distance apart, two dedicated public telephone lines form the link. Some geographically isolated sites use short-wave radio transmission. Operating experience of telemetry transmission over some distance suggests that the link is the weakest part of the installation

11.5. Fall back systems and standby facilities

A survey of existing practice shows wide variation between operators. In nearly all cases at least one stage of redundancy is introduced for critical equipment and in many cases two stages are provided.

Electrical supply is usually by two independent feeders or a ring main. With high voltage mains service, two transformers would complete such an installation, each transformer feeding a section of the busbar which is divided by an air circuit breaker.

Diesel engine driven standby generating plant, with or without automatic start up on mains failure, is general. The generating set is connected either permanently to the busbars via an interlocked circuit breaker or is of the portable type with plug-in facility.

The probability of failure of a diesel alternator set to start and run for two hours per demand is 0.043 per demand. Assuming that there is 1 demand in 2 years, the frequency of failure is approximately 1 in 46 years. This is not an acceptable level of reliability for emergency equipment. It therefore requires two standby generator sets for adequate security. The probability of simultaneous failure of two sets is 1 in 540 years. This assumes that there is no common cause failure—that is, an event or circumstance which affects both generating sets. Diesel fuel can be such a cause. Difficulties have been experienced with waxing during winter weather in cold climates, with fuel stratification due to long storage and bacterial growth in tanks. On this account the US Bureau of Reclamation favours the use of low-pressure gas engines for standby generating plant at spillway gate installations.

The problems which can be encountered with petrol engines used only intermittently are more severe than those experienced with diesel engines. While standby generation ensures a power supply in the event of mains failure, it does not provide for other electrical failures in an emergency such as a motor burn-out, failure of the busbar or motor starter. Some operating authorities consider that in a multi-gated installation, in the event of a flood there is sufficient time to exchange motors if a failure occurs. Others provide a spare motor as a standby. It is possible to arrange for the standby generating set either to plug into the busbar or to bypass the busbar and the individual motor starters as shown in Figure 11.8.

Standby equipment for servo-motors takes the form of a portable or mobile power pack which can be connected to the operating cylinder or cylinders by flexible high-pressure hoses and self-sealing couplings. For high reliability the portable power pack should have two diesel engines, each driving an oil pump.

Other emergency standby equipment for driving gate hoists are diesel engine driven hydrostatic transmissions which can be coupled to each gate in turn, either to the hoist gearbox or the hoist motor extension shaft. This is shown diagrammatically in Figure 11.9.

Figure 11.10 shows a diagram of a compressed air storage system with permanent air motors at each gate.
Battery-powered emergency drive systems are another means of making stored power instantly available. They can take the form of permanent DC motors (Figure 11.11). The alternative is to interpose an inverter to utilise the existing AC motors and starting equipment (Figure 11.12).

In a survey of emergency standby equipment and reliability analysis for spillway gate installations it was concluded that if two diesel generating sets are available and the starting and running of either set will suffice, then the probability that neither will start and run for two hours on demand is 0.0018, or the frequency of failure is 1 in 540 years. This does not take into consideration common-cause failures which can affect both sets, such as maintenance failure, dirty fuel or unavailability of fuel, fire, earthquake or unavailability of operating personnel. An assumption made in the calculation of failure rate is that each generating set is test run every two weeks, which has been shown to be the optimum on a statistical basis.

The reliability which can be obtained with a battery-powered emergency drive system is a frequency of failure of the order of 1 in 800 years.

Spillway gates not requiring external power in an emergency

Two types of gate can be opened under gravity:

'Gibb' gates, Victoria Dam, Sri Lanka[5]

The spillway gates of the Victoria Dam in Sri Lanka are counterbalanced so that they open under gravity and closure is effected by oil hydraulic cylinders. A description of the operation of these gates was given earlier in this chapter under 'Cascade Controls'.

There are a few spillway radial-gate installations where the gates are counterbalanced to open, but so far the only gates which operate automatically under gravity are the Gibb gates.

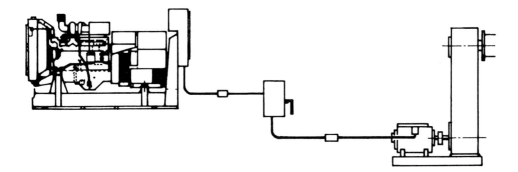

Figure 11.8. Standby diesel engine generator set.

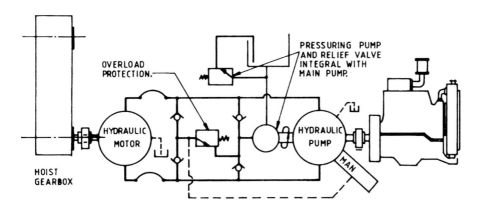

Figure 11.9. Standby diesel engine driven hydrostatic transmission

Bottom-hinged flap gates

Bottom-hinged flap gates have been used as spillway gates. A particular example is the Legadadi Dam in Ethiopia. Emergency lowering can be effected by manual operation of a directional control valve to port the annulus of the gate-operating cylinder to tank and to vent the cylinder side. The pressure of the reservoir water then lowers the gate.

An Australian Hydroelectric Authority operates automatic float-operated counterweighted gates at spillways. The gates are similar to those in Figure 2.11. The gates can be raised in the event of malfunction by shutting the displacer chamber outlet pipe.

The provision of manual winding of gates operated by electro-mechanical hoists is almost universal, although it is used as a last resort for opening larger gates because the time required can vary from six to twelve hours.

On oil hydraulic power packs for servo-motors, a hand pump is provided for operating the gate in an emergency. As in electro-mechanical drive

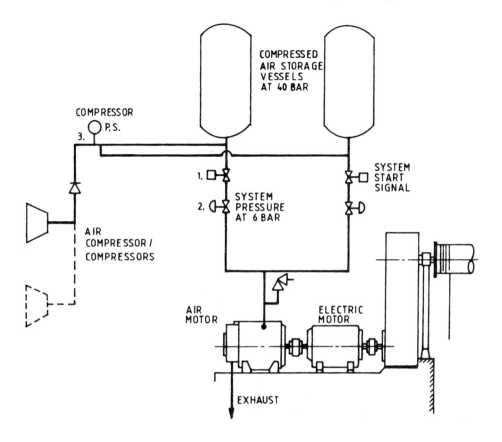

1. ON/OFF VALVES TO ISOLATE STORAGE FROM AIR MOTORS UNTIL
   THE SYSTEM IS REQUIRED TO OPERATE.

2. PRESSURE CONTROL VALVE TO MAINTAIN CORRECT PRESSURE
   TO AIR MOTOR.

3. PRESSURE SWITCH TO ACTIVATE COMPRESSORS TO MAINTAIN
   STORAGE PRESSURE IN VESSELS.

Figure 11.10. Schematic diagram of compressed air storage system providing air motor drive of the gate hoists

systems, it requires a long time to effect opening of a gate. This can be improved to a limited extent by fitting two pumps.

It is general practice to arrange hydraulic circuits so that in an emergency gates can be lowered by gravity (or in some cases opened) by manual operation of a control valve.

Alternative means of control

In all cases of automatic control, means are provided to revert to manual controls by an operator pressing 'open' and 'close' push buttons if the

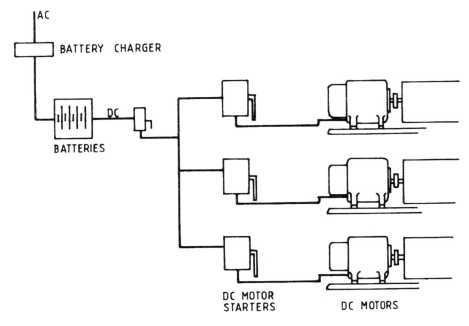

Figure 11.11. Schematic of battery power with DC motor drive

automatic controls fail to operate. These also serve to actuate the gates for testing, servicing and maintenance. The practice of testing spillway gates varies at different dams and with different operators[9].

Central computer operation for a cascade of dams or barrages is backed up by a local computer at each barrage, e.g. on the River Rhône[9]. At another

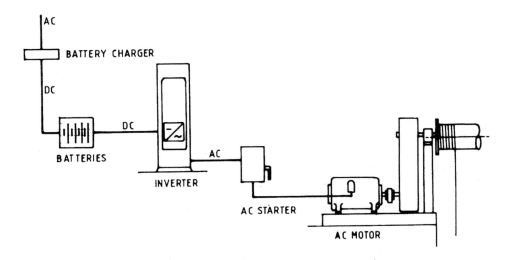

Figure 11.12. Schematic of battery power with AC motor drive

project the spillway gate control computer can, in the event of failure, transfer the control function to a second computer at the power station. At Kotmale Dam[8] an electro-mechanical, level-control system takes over if the computer control fails to maintain maximum retention level.

It appears that in all cases of computer control the system is made self-checking. At Kotmale[8] there is an independent checking and warning system over and above the checks carried out by the computer.

A difficulty arises in training operating personnel in manual operation of spillway gates which are normally automatically controlled. Their experience of dealing with flood events is limited and intermittent. Where a computer is provided such training could be by simulation. Although control strategies have been tested by computer simulation (Kotmale Dam), computers do not appear to have been used so far for operator training.

The practice of operating flood release from reservoirs, as well as staffing policy at dam outlets, is reviewed and tabulated in Combelles and Tinland[9].

11.6. Instrumentation

The two main parameters to be measured to control gates and valves are water level, which may involve both the control of upstream and downstream water levels, and gate or valve opening. In some cases direct flow measurement is also required, although this is more frequently deduced from calibration curves of water level and gate or valve opening.

11.6.1. Water-level measurement

Water level can be measured by electrodes, bubbler devices, ultrasonic sensors, pressure transducers or float gauges. The latter appear to be the most reliable instruments for limited distances of approximately 10 to 15 m. Pressure transducers tend to be more reliable for measurement of greater depth, but less accurate than precision bubbler devices which can be accurate to one half of one percent. At least two instruments are used for reservoir level measurement. Their results are averaged and compared and if they deviate by more than a predetermined amount, a warning is registered in the control room. The current trend is to use three instruments on a 'voting' basis; that is, the three signals are compared and if one deviates from the other two, its reading is ignored. This also actuates a warning signal.

11.6.2. Discharge from a gate

Discharge from a gate cannot be measured directly. It is usually inferred from measurement of upstream water level and gate opening.

Water level should be measured in a stilling well to avoid the false reading which would result from the velocity head in the sluiceway or its approaches. If means are provided for isolating the stilling well from the sluiceway approaches, the instruments can be calibrated and checked independently of the water level to be measured. A gauge board should be provided for checking the instruments or, if this is not practical, a piezo-electric water level gauge.

It is current practice to provide three pressure transmitters arranged on a 'voting' basis. The practice of providing three instruments of the float displacement kind requires appreciable space and increases cost. For reliable measurement at least two instruments should be used.

Discharge under gates is determined from the relationship of gate opening and head above the weir crest. This can be established from a model study. Most spillway sluiceways operate as if they are independent of the adjoining sluiceways, provided there is no significant super elevation. It is not therefore necessary to run the model for various combinations of gates and gate openings. These conditions arise if one gate is out of operation during a flood, due to maintenance, or if a gate fails to open due to malfunction.

To determine discharge over a gate, the total head above the weir crest and the weir coefficient must be known. For a valve, the head upstream of the valve and the valve rating curve are the measurement parameters. The coefficient of discharge for most valves is not constant throughout the range of valve openings and a rating curve is therefore required. However, in small closed conduits direct measurement of flow is usually possible.

11.6.3. Water level measurement instruments

Electrodes

Electrodes operate as 'on'–'off' devices. A number of electrodes can be used to provide cascade control of gates, that is where a gate or gates open in steps depending on water level. Each step corresponds to the setting of a limit switch.

For the control of level, two electrodes are used, set apart by the dead band, the upper and the lower limit of control. The upper electrode initiates raising of the gate in steps when it is covered. De-energising of the electrode causes the hoist motion to stop. Energising the lower electrode causes closure of the gate, also in steps.

The high-level electrode is usually duplicated with one acting as a standby to the other. An uppermost electrode is sometimes provided to warn when a danger level has been reached.

Electrodes can sometimes be energised by dripping water and can be protected against such inadvertent signals by a sheath of nylon or PVC.

Level gauges of the float actuation type

The float is linked by a tape to the instrument which displays level on a circular dial, usually with two hands like a clock. The float is steadied by guide wires. The measuring tape actuates a sprocket wheel which operates the hands through reduction gearing. Float level gauges are available with electrical analogue or digital measured value transmission and with a series of contacts for signalling high, low or intermediate levels. The measuring range can be up to 30 m. While the accuracy of the mechanical reading is about ±2 mm the electrical transmission accuracy and bias depends on the range of level that the instrument has to cover.

Pressure transmitters

Pressure transmitters are used for depth measurement. They are encapsulated, integrated, silicon, strain-gauge bridges. They are available from a range of pressure from zero up to 500 bar and even higher. For water level measurement, vented gauges are used with a conventional 4–20 mA range. Linearity and hysteresis are obtainable to ±0.1%. When good accuracy is required, it is advantageous to select a pressure transmitter which will only just exceed the required range of water level. Most pressure transmitters have a high overload capacity. For high reliability, three pressure transmitters are arranged on a 'voting' basis.

Precision pressure balances or bubbler devices

These instruments supply compressed air through a pneumatic tube to a point where the water level is to be measured. A small stream of air is allowed to bubble into the water at the discharge nozzle. The air pressure at the nozzle is proportional to the depth of the water at that point. This pressure is transmitted via the pneumatic tube and a service unit to the receiver, which applies a force proportional to the value to be measured to the balance beam system. When the equilibrium of the balance is disturbed by a change in the measured value, the displaced beam actuates a control contact. The servo-motor is actuated and moves a travelling mass until the equilibrium of the balance is re-established. The servo-motor and the travelling mass are connected via gearing to a digital display counter and analogue or digital switching and transmission units. The accuracy of bubbler devices can be ±0.25%.

Bubbler devices are arranged so that a blocked measuring nozzle can be cleaned by the application of full compressor pressure. Operationally, bubbler devices require more frequent checking and maintenance than float level gauges or pressure transmitters. They are frequently employed to measure head loss across a screen or where a wide range of water level has to be measured accurately. In practice, bubbler devices often fail to operate due to lack of maintenance or incorrect setting by inexperienced staff.

11.6.4. Gate or valve position measurement

This is carried out by indicating transducers. They can be of the type to measure angle of rotation or linear movement. Angular transducers convert angular deflection into a load-independent analogue signal. The input shaft is coupled to a reduction gear which drives the transducer via a friction clutch. A visual angle of position indicator is frequently incorporated in the instrument. Some instruments are available with an integral second angular transducer to give a greater accuracy over a limited range. When the range of displacement of one transducer is exceeded, the second transducer is coupled in and rotates to the end of the required range. Limit switches or changeover contacts form part of some instruments.

Angular transducers are also used to measure linear movement by converting linear displacement to angular motion by passing a cable over a pulley or by linkage.

References

1. Lewin, J (1985): The Control of Spillway Gates During Floods, *2nd International Conference on Hydraulic Aspects of Flood and Flood Control, Cambridge*, B.H.R.A., Fluid Engineering.
2. Lewin, J; Denham, H (1983): An Adaptive Control System for Flood Routing through a Reservoir, *1st International Conference on Hydraulic Aspects of Flood and Flood Control, London*, B.H.R.A., Fluid Engineering.
3. Anon (1976): *Flood Control by Reservoirs*, Chapter 6, Spillway Operation, Section 6. Considerations for Spillway Operation. Hydrologic Eng. Centre, U.S. Army Corps of Engineers, Feb.
4. Evans, T E; Halifax, P J; Floyd, D S (1983): A Real Time Computer Operated Model developed for the River Medway Flood Storage Scheme, *1st International Conference on Hydraulic Aspects of Flood and Flood Control, London*, B.H.R.A., Fluid Engineering.
5. Back, P A A; Wilden, D L (1988): Automatic Flood Routing at Victoria Dam, Sri Lanka, *Commission Internationale des Grands Barrages, 16th Congress, San Francisco*, Q63, R52.
6. Di Stefano, J J; Stubbard, A R; Williams, I J (1976): *Feedback and Control System*, McGraw-Hill.
7. Gosschalk, E M; Longman, A D (1985): Sri Lanka's Kotmale Hydro Project, *International Water Power and Dam Construction*, Mar.
8. Lewin, J (1987): The Spillway Gates and Bottom Outlet of Kotmale Dam, *International Water Power and Dam Construction*, Aug.
9. Combelles, J; Tinland, J M (1984): *Operation of Hydraulic Structures of Dams*, Commission Internationale des Grands Barrages, Bulletin 49 Appendix 1, Caderousse on the Rhône River, France.

12
Extreme environmental factors

12.1. Ice

Criteria for the design of gates under ice conditions in Northern Europe are given in DIN 19704[1], where different empirical rules apply to inland and estuarial conditions.

12.1.1. Empirical criteria for inland conditions

In the design of skin plates and their associated stiffening members, as well as for the main girder, it is assumed that the formal triangular hydrostatic distribution of water pressure at a depth of 1 m is replaced by:

- An even surface pressure of 30 kN/m^2 where the ice formation is 300 mm
- An even surface pressure of 20 kN/m^2 in waters with moderate ice formation up to 300 mm thick.

12.1.2. Empirical criteria for estuarial conditions

In the design of skin plates and their associated stiffening members, the following loads shall be assumed over and above the hydrostatic loads within 0.5 m above and below water level:

- An even surface pressure of 100 kN/m^2 when severe ice formation is present and ice displacement occurs.
- In conditions of moderate ice formation, an even surface pressure of 30 kN/m^2 should be used.

For main girders additional loads shall be applied to the node points level with the water surface and amounting to:

- A distributed load of 350 kN/m in severe ice conditions.
- A distributed load of 100 kN/m in conditions of moderate ice formation.

The design of underflow gates should be checked for a distributed load of

30 kN/m at the gate lip. Ice forming within the gate structure should also be taken into account.

Starosolsky[2] gives extensive information on ice formation and provides some data which can be used to arrive at a more rational basis of design. Otsubo[3] and Johansson[4] give examples of the effect of severe ice formation on gates and some means that have been tried to mitigate it.

Under winter conditions of light frost, the side seals of gates can freeze to their contact face. An attempt to operate a gate under these conditions could result in tearing of the seal.

Where there is an operational risk of the side seals freezing to their contact face, either because moisture is trapped between the seal and its contact face or because there is leakage past the seal, the side staunching has to be heated when air frost occurs. Lintel seals will also have to be protected against freezing.

Heating is effected by electrical resistance cables, usually of the mineral-insulated, stainless steel sheathed kind (Figure 12.1). An alternative method is by circulating hot oil. Heating provision was made at the Tees Barrage in the UK where it was provided for some of the bottom-hinged barrage gates in their closed position. The seal contact plates of the large vertical-lift gate of the Barking Barrier also incorporate provision for heating. Heating under frost conditions of side-seal contact plates for radial gates is extended throughout the length of travel of the side seals. Heating cables have to be insulated to prevent undue conduction of heat to the flume walls.

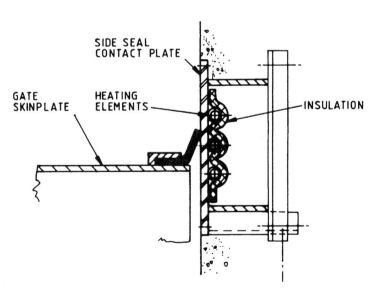

Figure 12.1. Heating cables for side seal contact face

In England an approximate heating load of 1 kW/m^2 is considered adequate when the gate may have to be moved under conditions of air frost. For central European conditions an approximate heating load for side-seal contact faces is 1.5 kW/m^2. Under severe frost, when the upstream and the downstream water is subject to ice formation to some depth, it is not practical to move a gate and it is unlikely to be required.

12.2. Seismic effects

There is little published on the effect of earthquakes on gates. At dam crests and outlet towers, seismic shocks will be considerably magnified. The most frequently recorded effects relate to travelling gantries for hoisting gates or for placing stoplogs. These have jumped rails due to earthquake shocks. Where an acceleration factor due to possible seismic effects has to be taken into consideration, the portal structure of a crane has to be designed for the horizontal and vertical impact which can be transmitted through the crane rails. The flanges of rail wheels have to be dimensioned to withstand the additional horizontal thrust which will be transmitted through the wheels to the end carriages.

To prevent derailment, rail clamps are provided and the rail fixings designed for uplift. Rail clamps are also provided to stabilise travelling gantries against overturning or inadvertent travelling when high wind loads occur.

Gates suspended by ropes or a series of operating rods can be subject to accelerating forces in excess of the force applied to the hoist.

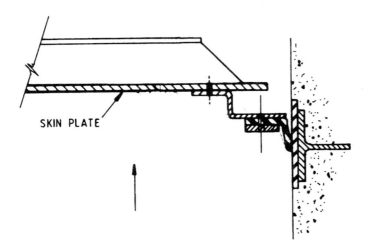

SKIN PLATE

Figure 12.2. Side seal mounting of a radial gate to provide a collapse zone in the event of a transverse movement of a sluiceway pier due to an earthquake

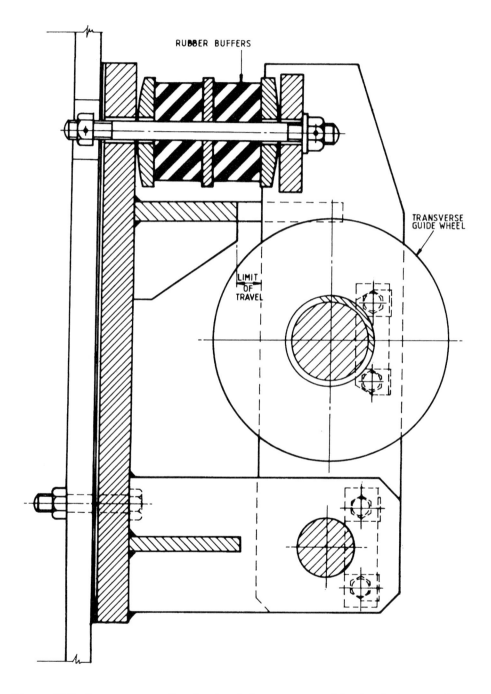

Figure 12.3. Arrangement of spring loaded transverse guidewheel of a tunnel gate to absorb an earthquake shock

The transverse movement and possible displacement of sluiceway piers due to earthquakes is a possibility. Radial gates, due to their close tolerance to effect a side-seal, can become wedged or suffer local buckling. Under these conditions the side-seal mounting is arranged so that the seal mounting section will deflect or buckle under severe impact as shown in Figure 12.2. It is more usual to arrange the seal mounting bracket upstream of the skin plate to prevent debris accumulation in the pocket formed between the seal-mounting plate and the skin plate. The practical difficulty which arises due to such an arrangement is to bridge the side and the sill seals at the junction between the two seals. This is done by a rubber block, which can be a source of leakage.

In radial gates, earthquake forces in the form of vertical and horizontal acceleration will be transmitted through the pivot bearings of the gate arms, the sill beam and the hoisting chains or ropes. In vertical-lift roller or slide gates, horizontal accelerating forces will be applied through the transverse guide slippers or guide wheels. Figure 12.3 shows an arrangement of spring-loaded transverse guide wheels to absorb shock. This arrangement is also used on stoplogs and draft-tube gates when they are placed under balanced pressure, in order to locate them close to the sealing faces for good initial sealing.

National codes of structural design based on permissible stresses are being revised to the load factor method. This is also the case where a national standard for hydraulic gates exists, such as DIN 19704[1]. When designing gates and valves to withstand imposed loads due to earthquake shock, the load factor method is more appropriate.

References

1. DIN 19704 (1976): *Hydraulic Steel Structures: Criteria for Design and Calculation.*
2. Starosolsky, O (1985): In *Developments in Hydraulic Engineering*—3, editor Novak, P, Chapter 5, Elsevier Science Publishers, p. 175–219.
3. Otsubo, K (1959): Ice Problems of Gates at Hydro-Electric Plant in Northern Districts of Japan, *8th I.A.H.R. Congress, Montreal*, Vol. III, p. 2–S1–1 to 2–S1–2.
4. Johansson, H (1959): Ice Problems relating to Dam Gates, *8th I.A.H.R. Congress, Montreal*, Vol. lll, p. 27–S1–1 to 27–S1–3.

13
Materials and protection

13.1. Materials

Most gates are fabricated in low-carbon structural steels. Except in high-head gates, the higher tensile stress grades are rarely employed because deflection often becomes the critical design parameter so that sealing faces do not open under load. In gates which are subject to ice formation (see Section 12.1), low-temperature structural steels are used to prevent brittle fracture.

In fluidways and valves where high-velocity flow is experienced, and where there is a risk of cavitation, stainless steel is selected. The same material is used for linings downstream of gates in high-head tunnels where the boundary layers have not developed sufficiently to protect the walls from high-velocity flows. Seal contact faces and guide roller paths are also constructed in stainless steel.

When considering corrosion resistance, the total amount of the nickel and chromium in the steel is an indication of the degree of corrosion resistance. Therefore, the high nickel and chromium austenitic steels are generally the best.

Stainless steels are often welded to low-carbon structural steels. To prevent carbon migration at the welds, stabilised austenitic chromium nickel steels must be used. The steel is stabilised by the addition of titanium. Under some conditions, unstabilised stainless steels can be welded to low-carbon steels, using special welding rods.

Stainless steel clad, carbon steels are used to effect cost reductions. The layers of stainless steel are rolled onto the plate to produce an integral plate. It is desirable that the cladding be at least 1.0 mm thick. Welding of a clad steel to a carbon steel is possible but requires a special technique. The protection of cut faces and of bolt holes in clad steels presents problems. The cost advantage of clad steels tends to fluctuate.

A limited number of cases of electrolytic underwater corrosion have been reported, due to the proximity of stainless and carbon steels but, in general, operating experience in inland waters has been favourable.

In saline waters, nickel copper alloy (Monel metal) is used, and also in fresh water where parts are in contact with brass or bronze. Nickel copper alloy is available in plate and sheet form, rod and bar. It is more expensive than stainless steel. It is closer to the brass and bronze alloys in the galvanic series and its corrosion resistance, especially in saline waters, is superior to that of stainless steels.

Guide rollers, shafts and pins, which are permanently or occasionally submerged in water, are selected in high chromium, ferritic stainless steel, in free machining austenitic chromium nickel steel or nickel copper alloy. Their use in connection with bearing materials of leaded bronze is general.

Martensitic stainless steels can be suitable for some applications but should be used with caution for heavily-loaded rotating parts, as some have a relatively low fatigue limit.

Bolts and nuts, particularly where they may have to be removed for maintenance or replacement, should be in stainless steel or Monel. Bronze or brass bolting is not suitable.

The usual range of engineering materials from ductile iron castings to alloy steel castings, from medium carbon to high tensile alloy steels, are used. Their selection and application is similar to general engineering practice except that factors of safety are generally higher. Grey iron castings are not, as a rule, used for stress carrying components in gate installations because they are liable to brittle fracture.

13.2. Paint systems

The corrosion protection of hydraulic structures, particularly gates, is of the utmost importance to the operating authority. To carry out maintenance painting of a gate the sluiceway has to be stoplogged, pumped dry and in many cases a protective shelter has to be provided over the area to be painted. Apart from the cost of maintenance painting, the conditions of carrying out the work are rarely good.

The selection of the most appropriate paint system is difficult. There are a number of books[1,2] and papers[3,4] which give guidance, as well as the British Standard 5493[5]. The sequence of selection in the BS is given by a table, but to an engineer who is not a specialist in the subject, it does not assist in choosing from many combinations of acceptable systems.

A frequent practice is to blast clean to Swedish Standard, Sa2½, to apply a wash coat and repeated applications of epoxy coal tar to 350 mm minimum dry film thickness. When applying the epoxy coal tar coating under normal ambient temperatures and normal curing, a polyamide-cured epoxy is used. Under low ambient temperatures, when quick curing is necessary, an isocyanate-cured epoxy coal tar is applied.

Epoxy coal tar paints are available in a restricted range of colours. This can present a difficulty when painting spillway gates in the tropics. It is desirable to reduce the solar heat gain on the downstream side of gates by using a white or aluminium colour coating. The differential expansion of a spillway gate due to the heating of one face by the sun, and a lower temperature due to reservoir water on the upstream side, can result in leakage at the sill seal.

Up to some years ago, metal-sprayed and thick paint coatings were widely used on bridges. There is no literature on the effectiveness and durability of this protective treatment for gates.

The large radial automatic gate at the Pulteney sluices in the City of Bath was zinc sprayed to 250 mm without subsequent painting. The protection was effective for approximately 25 years. The difficulty of ensuring adequate and consistent thickness of a sprayed coat at corners and sharp re-entrant angles does not recommend a repetition of this treatment.

Most gates are assembled with some bolted connections. The mating surfaces must be protected from corrosion, even if high-strength, friction-grip bolts are used. Joints which have been coated and are assembled with friction-grip bolts have to be derated. The CIRIA[6,7] and the Transport and Road Research Laboratory[8] reports offer guidance. Some gate designers favour seal welding of bolted connections.

High-head gates often incorporate enclosed sections. The delta configuration of the lip of a slide or roller gate is an example. It is possible for oxygen and moisture to penetrate into such enclosed sections. Moisture penetrates by differences in vapour pressure, and as a result of changes in temperature it can condense and cause local corrosion. Bolt holes in enclosed sections should be avoided, and welds, which must be continuous, should be tested to avoid discontinuities and pinholes.

Since paint selection, preparation for painting, the testing of painted surfaces and repair of damaged coating form a specialist subject, the preceding notes are only intended to mention some factors pertaining to hydraulic structures and to provide reference to literature for further reading.

13.3. Cathodic protection

Cathodic protection is the technique of reducing the corrosion rate of an immersed metallic structure by making the steady state or corrosion electrical potential of the metal sufficiently more electronegative. The thermodynamic considerations are dealt with by Shreir[9], Pearson[10] and others.

When two dissimilar metals are electrically connected and immersed in an electrolyte, which can be fresh or salt water, there is a current flow through the electrolyte and the metal so that anions enter the solution from the anode, and at the same time electrons move from the anode to the cathode via the metallic connection. The rate of corrosion protection depends on the amount of current flowing, and this depends on the electro-magnetic force (e.m.f.) and various ohmic and non-ohmic resistances in the circuit.

The e.m.f. may be provided by a metal which is more electronegative than the metal to be protected (sacrificial protection) or by an external e.m.f. and an auxiliary anode (impressed current protection).

Cathodic protection of gates in sea water or estuarial locations is mainly by the use of sacrificial anodes of zinc[11], magnesium or aluminium. Impressed-current cathodic protection is used when corrosion conditions are severe and where inspection and remedial work during the lifetime of the structure is impossible or not practical. Cathodic protection is not effective in the splash zone of a gate and in a tidal application will afford only reduced protection in the upper tidal zones.

The current density required for steel for adequate cathodic protection in moving fresh water is 55–65 mA/m^2. In stilling basins, where the water is highly turbulent and can contain dissolved oxygen, the range is 55–165 mA/m^2. In sea water the current density for cathodic protection is within the range of 55–300 mA/m^2, whereas in highly-polluted estuarine water 600–2000 mA/m^2 may be required.

Magnesium is probably the most widely used of the sacrificial anode materials as the high current yield ensures maximum current distribution. The addition of aluminium to magnesium reduces self corrosion, but minor alloying elements such as copper, nickel and iron can significantly increase self corrosion and decrease the efficiency of magnesium as a sacrificial anode. Alloying elements are therefore controlled within limits in magnesium anodes.

Current output is related to the composition of the anodes, surface area and shape, while the life is dependent on the ratio of surface area to weight, together with the current demand of the water at the gate location.

Although the principles of cathodic protection are essentially simple, the practical application to the protection of steel structures, such as gates, immersed in water appears to be more of an art than a science.

Cathodic protection applied to a structure, and particularly when it is applied only to elements of a structure, can present a danger to adjacent unprotected structures or parts.

Cathodic protection can be provided to prevent cavitation damage. It requires high current densities in order that the hydrogen freely evolved from the protected metal can act as a gas cushion between the collapsing vapour cavities and the metal surfaces. This makes it impractical to protect more than a limited surface area. Where cavitation cannot be avoided, such as the area downstream of a high-head tunnel gate, it is more economical to provide a replacement liner.

References

1. Hudson, J C (1940): *The Corrosion of Iron and Steel*, Chapman and Hall.
2. Evans, L I R (1960): *The Corrosion and Oxidation of Metals*, Edward Arnold Ltd, Chapter 13.
3. CIRIA (1982): *Painting Steelwork*, editor Haigh, I P.

4. HMSO (1971): *Report of the Committee on Corrosion and Protection*.
5. British Standard 5493:1977, *Code of Practice for Protective Coating of Iron and Steel Structures against corrosion*.
6. CIRIA (1969): *Protection of Steel Faying Surfaces*, editor Day K J, Interim Research Report.
7. CIRIA (1980): *Design Guidance Notes for Friction Grip Bolted Connections*, editor Cheal, B D, Technical Note 98.
8. Black, W; Moss, D S (1968): *High Strength Friction Grip Bolts—Slip Factors and Protected Faying Surfaces*, Transport and Road Research Laboratory, Report LR 153.
9. Shrier, L L (1963): *Corrosion*, section 11, Cathodic Protection, George Newness Ltd.
10. Pearson, J M (1955): Fundamentals of Cathodic Protection, in Section Vll, Corrosion Protection, *The Corrosion Handbook*, editor Uhlig, H H, John Wiley & Sons Inc./Chapman and Hall Ltd.
11. Day, K J (1977): Protective Treatment in *Proc. I.C.E., Conference Thames Barrier Design*, 5th Oct, paper 16.

14
Model studies

Numerous model studies of gates and gate installations have been carried out. Many of them relate to a particular project and were undertaken to obtain specific numerical results. These have limited validity and show only certain relationships of observed hydraulic parameters within the range of the experiments undertaken. Some studies have explored problems which are encountered in other installations and by stating the physical principles shown by an empirical function can result in extrapolation of the experimental data or formulation of general guidelines for other similar cases of flow. Study of papers can therefore add to experience and enhance understanding of fluid dynamic behaviour.

14.1. Froude scale models

The vast majority of models are to Froude scale. This represents the condition of dynamic similarity for flow in model and prototype exclusively governed by gravity. It cannot be used to determine other forces such as frictional resistance of a viscous liquid, capillary forces, the forces of volumetric elasticity and cavitation phenomena.

To obtain similarity between model and prototype for flow conditions where inertia and gravitational forces are dominant, the Froude number F_n of the model and the prototype must be the same.

$$F_n = V/\sqrt{(gd)}$$

where V = flow velocity
g = gravitational constant
d = depth/length

Thus at a model scale of 1 in S:

$$\text{Flow} = Q_m = Q_p/S^{2.5}$$

$$\text{Velocity} = V_\text{m} = V_\text{p}/S^{0.5}$$

$$\text{Time} = T_\text{m} = T_\text{p}/S^{0.5}$$

the suffix m denoting model, and p prototype conditions.

For viscous forces it can be shown by dimensional analysis that the Reynolds number R_n in the model and prototype should be the same. Both the Reynolds and the Froude numbers for a model and prototype cannot be made equal. Any difference in the Reynolds number is not particularly important as long as both model and prototype have high values, above 100,000, and similar roughness to diameter ratios. Under these conditions the headloss is a common function of the square of the velocity in both model and prototype. If the reduced Reynolds number of the model approaches the point of transition of turbulent to laminar flow, the laminar flow could occur in the model but turbulent flow would occur in the prototype. This must be avoided, and consequently a minimum operable Reynolds number has to be chosen. Increasing the velocity to improve Reynolds number correlation is often used to test for safety margins that have been eroded by these non-scale effects. A full discussion of the theory of similarity is given in Novak and Cábelka[1].

14.2. Two-phase flow problems

A number of hydrodynamic problems encountered in overflow and tunnel gates involve two-phase flow. Air may be entrained at intakes to conduits or drop shafts and, due to reduced pressure caused by high velocity flow or flow transition phenomena, is liberated in the tunnel. In steady flow in open channels, air entrainment depends on flow velocity, generally at about 6 m/s and higher flow velocities. A frequent cause of air entrainment in hydraulic structures is steady flow transition, such as a hydraulic jump in closed conduits, the transition phenomenon of a jet into pressure flow. The same effect occurs in overflow gates due to the suction effect on the inside of the nappe. Measurement of two-phase flow is possible and is discussed by Novak and Cábelka in section 4.7 of their book, but the quantitative results are valid only under prototype conditions

14.3. Two- and three-dimensional models to Froude scale

Hydraulic models of gates and gate installations can be divided into three categories, although such a division is arbitrary and categories overlap.

The two-dimensional model can be constructed in a flume and is intended for the study of gate characteristics and to verify the design of an associated stilling basin or a weir crest. The objectives of such a model may range from determining the discharge characteristics of a radial gate, the hydraulic downpull forces acting on a vertical-lift gate in a tunnel, to studying the effect of interaction of an operating and a guard gate in a conduit. An example in Figure 14.1 shows the model study of the spillway gates of the Kotmale Dam in Sri Lanka[2]. The objectives were to determine the gate characteristics throughout the range of gate openings, but particularly at small gate apertures. These were required to programme the gate discharge at the onset of a flood to increase gradually the downstream river flow to safeguard and warn river users. Another requirement was to operate the gates to attenuate return period floods of up to 100 years. The model was also used to measure sub-atmospheric pressures on the upper part of the spillway, to determine pressure variations across the weir crest and below.

The second category of the model is the three-dimensional approach flow, which includes a significant section of the downstream geometry. An example of this type of model is the River Medway flood relief scheme[3] where the pattern of flow from the storage reservoir into the sluiceways was a part of the investigation. The model of the tunnel gates in the bottom outlet of the Mrica Hydropower Scheme[4] is another three-dimensional model. The conduits from the two gates split the approach channel and reunite downstream of the gate installation. The hydraulics of splitting the flow, operating one gate only, or two gates with different discharge, and

Figure 14.1. Model study of the spillway gates of the Kotmale dam

uniting the flow downstream of the gates, were the major reasons for the model study.

14.4. Models for investigating vibration problems

The third category of model study is the investigation of existing or potential problems of gate vibration or cavitation. Some problems of flow-induced structural vibrations can be investigated by reproducing in the model a single degree or multiple degrees of freedom[5,6]. However most studies require that the model be constructed with an overall reproduction of elasticity. This provides a check of the design. It involves the use of special plastics. The design and construction of the model is lengthy and expensive.

The scaling criteria, which must be satisfied in constructing a hydro elastic model, are given in Kolkman[7], Appendix C and Haszpra[8]. Most of the models of hydroelastic similarity for studying potential gate vibration problems were investigated at Delft Hydraulics, such as the Hagestein visor gates[9], the radial gates of the Haringvliet sluices and the vertical-lift gates of the storm-surge barrier across the Eastern Scheldt (Figure 14.2). In the United Kingdom the rising-sector gates of the Thames Barrier were the subject of hydrodynamic load and vibration studies[12,13]. Studies to compare model and prototype results[14,15] have been undertaken. These, taken together with the success of designs which have been tested by models of hydroelastic similitude, justify confidence in vibration models.

Some actual or potential vibration problems can be investigated by constructing a model only to Froude scale. This requires observation of the flow conditions, and experience to judge whether these are likely to cause gate vibration. Models which reproduce possible conditions of flow separation and reattachment at bottom sections of gates can be studied in this manner.

Cavitation can be an important cause of dynamic load as well as causing significant loss of material. Gates have to be designed to be cavitation free, or to be subjected only temporarily to a low degree of cavitation. To satisfy the theoretical requirement, model research must be such that model and prototype vapour pressures occur at equivalent locations. The pressure at any other location in the flow, with reference to the pressure in the critical location, is related to ρV^2 (where ρ is the density of water and V is the reference velocity). Therefore the criterion is that:

$(\rho - \rho_{\text{vapour}})/\rho V^2$ (the Thoma number) is correctly reproduced.

where ρ = water pressure

    ρ_{vapour} = vapour pressure of air

This cannot easily be done and requires a special facility[16]. Such a test facility was developed by the Snowy Mountains Engineering Corporation in its Fluid Mechanics Laboratory[16]. Small-scale models have been used successfully at Imperial College, London, to indicate likely patterns of cavitation and even of cavitation erosion for the elements of a large structure[17,18].

Figure 14.2. Model of hydroelastic similarity of the gates for the storm surge barrier of the Eastern Scheldt

References

1. Novak, P; Cábelka, J (1981): *Models in Hydraulic Engineering—Physical Principles and Design Applications*, Pitman Advanced Publishing Program.

2. Milan, D; Habraken, P (1984): *Kotmale, Report on Spillway Radial Gates, Model Tests*, General Technical Services, Lyon. Not published.

3. Palmer, M H (1979): *Hydraulic Model Study of the River Medway Flood Relief Scheme Control Structure*, B.H.R.A., Report RR 1572.

4. Bruce, B A; Crow, D A (1984): *Mrica Hydroelectric Project: Hydraulic Model Study of the Drawdown Culvert Control Structure*, B.H.R.A., report RR 2325, Nov.

5. Abelev, A S (1959): Investigations of the Total Pulsating Hydrodynamic Load Acting on Bottom Outlet Sliding Gates and its Scale modelling, *8th I.A.H.R. Congress, Montreal*, paper A10.

6. Abelev, A S (1963): Pulsations of Hydrodynamic Loads Acting on Bottom Gates of Hydraulic Structures and their Calculation Methods, *10th I.A.H.R. Congress, London.*

7. Kolkman, P A (1976): *Flow-induced Gate Vibrations*, Delft Hydraulics Laboratory, Publication No. 164.

8. Haszpra, O (1979): *Modelling Hydroelastic Vibrations*, Pitman Publishing.

9. Kolkman, P A (1959): Vibration Tests in a Model of a Weir with Elastic Similarity on Froude Scale, *8th I.A.H.R. Congress, Montreal*, paper A29.

10. Kolkman, P A (1979): *Development of Vibration-free Gate Design: Learning from Experience and Theory*, Delft Hydraulics Laboratory, publication No. 219, Nov; also *The Haringvliet Sluices*, Rijkswaterstaat Communications, 11th Nov 1970.

11. De Jong, R J; Korthof, R M; Perdijk, H W R (1982): Response Studies of the Storm Surge Barrier of the Eastern Scheldt, *Int. Conference on Flow Induced Vibrations in Fluid Engineering*, Reading, Sep, paper A3.

12. Crow, D A; King, R; Prosser, H J (1977): Hydraulic Model Studies of the Rising-sector Gate; Hydrodynamic Loads and Vibration Studies, *Int. Conference on Thames Barrier Design*, London, Oct, I.C.E., London, 1978.

13. Hardwick, J D (1977): Hydraulic Model Studies of the Rising-sector Gate conducted at Imperial College, *Int. Conference on Thames Barrier Design*, London, Oct, I.C.E., London, 1978.

14. Geleedst, M; Kolkman, P A (1963): Comparison of Measurements on the Prototype and the Elastically Similar Model of the Hagestein Weir, *10th I.A.H.R. Congress, London*, paper 3.21.

15. Geleedst, M; Kolkman, P A (1965): Comparative Vibration measurements on the Prototype and the Elastically Similar Model of the Hagestein Weir under Flow Conditions, *11th I.A.H.R. Congress, Leningrad*, paper 4.7.

16. Lesleighter, E J; Harrison, R D (1981): Development of a Cavitation Test Facility, *Institution of Engineers, Australia Conference*, Mar.

17. Kenn, M J; Garrod, A D (1981): Cavitation Damage and the Tarbela Tunnel Collapse of 1974, *Proc. I.C.E.*, Part 1, 70, Feb: Discussion *Proc. I.C.E.*, Part 1, 1982, 70, Nov.

18. Kenn, M J (1983): Cavitation and Cavitation Damage in Concrete Structures, *Proc. 6th Int. Conference on Erosion by Liquid and Solid Impact, Cambridge*, Sep.

Index